PHOSPHORUS - RECOVERY AND RECYCLING

Edited by **Tao Zhang**

Phosphorus - Recovery and Recycling
http://dx.doi.org/10.5772/intechopen.74920
Edited by Tao Zhang

Contributors

Ming-Der Su, Jia-Syun Lu, Ming-Chung Yang, Lucy Ngatia, Robert Taylor, Sayma Samreen, Sharba Kausar, Harada Hiroyuki, Inoue Kastutoshi, Agnieszka Saeid, Tao Zhang

First published in London, United Kingdom, 2019 by IntechOpen
IntechOpen is the global imprint of INTECHOPEN LIMITED, registered in England and Wales, registration number: 11086078, The Shard, 25th floor, 32 London Bridge Street
London, SE19SG – United Kingdom
Printed in Croatia

British Library Cataloguing-in-Publication Data
A catalogue record for this book is available from the British Library

Additional hard and PDF copies can be obtained from orders@intechopen.com

Phosphorus - Recovery and Recycling, Edited by Tao Zhang
p. cm.
Print ISBN 978-1-83881-021-4
Online ISBN 978-1-83881-022-1
eBook (PDF) ISBN 978-1-83881-023-8

For EU product safety concerns:
IN TECH d.o.o., Prolaz Marije Krucifikse Kozulić 3, 51000 Rijeka, Croatia,
info@intechopen.com or visit our website at intechopen.com.

Meet the editor

Tao Zhang is an associate professor and doctoral supervisor of the China Agricultural University. So far, he has served as a member of the Recycling Economy Committee and the Water Treatment and Reuse Committee of the Chinese Society of Environmental Science. His research area is the direction of waste disposal and resource utilization. He has chaired four key projects, including the synergistic mechanism of the microwave effect on phosphorus crystallization of livestock wastewater, the control of hazardous substances in manure and biogas slurry, and the key technologies for the safe utilization of farm land, etc. He has published more than 40 academic papers (18 SCI papers as the first author or corresponding author), and applied for 20 invention patents (14 have been authorized).

Contents

Preface

In the past few years, a number of physical, chemical, and biological methods have been used to recycle phosphorus more effectively. However, most of the recovery processes are developed for industrial wastewater and municipal wastewater and only a few recover phosphorus from fertilizers and biogas slurry.

In this book, the raw materials of phosphate fertilizer, the eutrophication status and mitigation strategies of phosphorus, and the critical key link of the phosphorus cycle—microbial solubilization, effective methods of phosphorus recovery, and the bonding properties of phosphorus—are illustrated. In addition, the author has summarized new discoveries on the basis of previous studies, and also established the guiding ideology of establishing a partnership and strategic framework for phosphorus recovery and utilization.

Tao Zhang
China Agricultural University
Beijing, China

Section 1

Introduction

Chapter 1

Introductory Chapter: The Overview of Phosphorous Recovery

Huanhuan Li and Tao Zhang

Additional information is available at the end of the chapter

http://dx.doi.org/10.5772/intechopen.85967

1. Introduction

Phosphorus (P) is a finite, non-substitutable, non-renewable, and geographically restricted resource. Substantial interest in P availability was sparked [1]. Some researchers demonstrated sufficient availability to sustain production beyond the twenty-first century, or a maximum occurring late twenty-first century [2–4]. But there are no financial incentives to support mineral resource inspections worldwide without the exploitation of phosphate deposits. The anthropogenic influences on this critical resource are likely to bring about a number of challenges to P sustainability. Advances in technology, public health, and food production over the last couple of centuries have fundamentally interrupted the natural global P cycle. Phosphate deposits have been mined to supply human production, which generated a mostly one-way flow of P from mines to farms to surface waters, ultimately impairing freshwater and coastal waters environment and function [5]. Rapid increases in human population and the subsequent need for high agricultural productivity have led to substantial increases in fertilizer use [6]. The P used as fertilizer consumes more than 80% of the P resources [7]. P is simultaneously an important non-renewable agricultural nutrient and an environmental pollutant [8].

The use of P resources in different countries or regions is different. But the basic P flow is as follows: phosphates are extracted from phosphate rock, passing through crops, animals and human, and ending up in landfill or emission into rivers and the sea (if sewage or animal wastes are not adequately treated) (**Figure 1**). The P cycle is extremely inefficient and wasteful. P loss from wastewater into natural water bodies accounts for about 10% of inorganic P source fertilizers in arable soils globally [9]. Eutrophication is the consequence of human behavior changing the natural P cycle. Researchers and practitioners in multiple fields have responded with numerous strategies to reduce P loading to aquatic ecosystems in an era of eutrophication control [10]. P has become the principal contributor and limiting factor to water eutrophication.

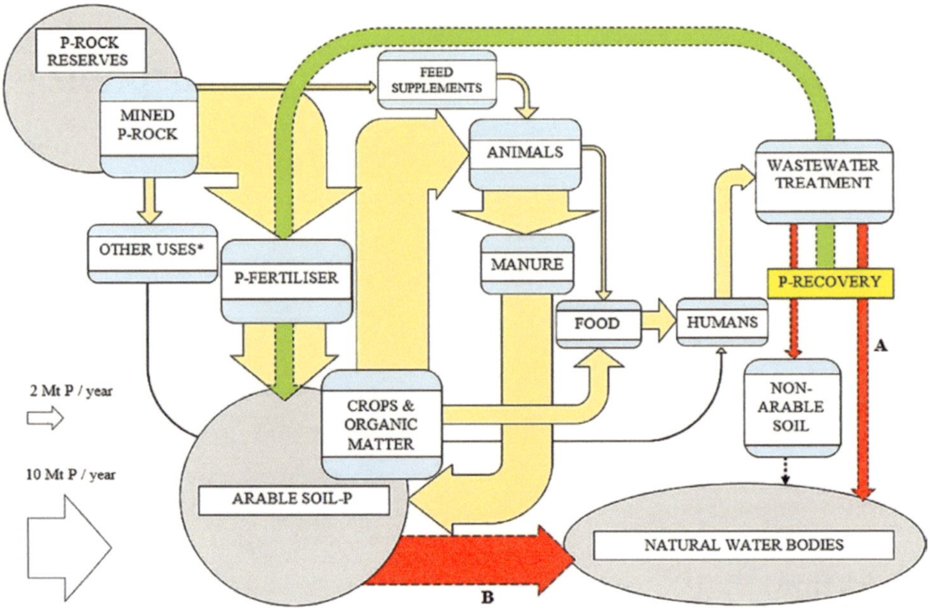

Figure 1. Diagram of key global P flows (the widths of the arrows semiquantitatively represents figures reported by Cordell [1] in million tons (Mt) of P per year Point "A" denotes the flow of P contained in treated or untreated sewage to natural water bodies and represents approximately 1.5 Mt P/year. Point "B" denotes the flow of P contained in erosion losses and is about 8 Mt P/year. *Other uses include industrial uses such as the production of some detergents) [9].

P, as a resource in fertilizer production or as a pollutant in wastewater treatment, is often simultaneous. The removed P has the potential to be reused as a substitute for mined P fertilizer [11]. With different fractions of P in different water or waste, direct use as fertilizer could not be the sustainable and efficient way and an increasing number of physical, chemical, or biological methods with various functions were designed to recover and reuse available P in the past few years: struvite crystallization, P adsorption, anaerobic digestion, membrane concentration technology, and integration technologies. Most of the P recovery processes are developed to be applied on industrial and municipal wastewater and only a few techniques are developed for P recovery from manure and digestate [12]. More sustainable techniques, such as P recovery techniques for both solid and liquid wastes, are important to maintain the P cycle in modern human society. Sustainable P use has been largely driven by pollution concerns and the shortage of P resources over the past few decades, which P will need to be recovered for productive reuse as a fertilizer to replace increasingly scarce P resource. But for P recycling and reusing, technological advances alone are not enough. There is no single solution to achieving a P-secure future. Partnerships and strategic frameworks to stimulate and support the development of renewable P is an urgent need to establish.

Acknowledgements

The work was supported by a grant from the National Key Technology Research and Development Program of China (Nos. 2017YFD0800202 and 2016YFD0501404-6).

Author details

Huanhuan Li and Tao Zhang*

*Address all correspondence to: taozhang@cau.edu.cn

Biomass Engineering Center, Beijing Key Laboratory of Farmland Soil Pollution Prevention and Remediation, Key Laboratory of Plant-Soil Interactions of Ministry of Education, College of Resources and Environmental Sciences, China Agricultural University, Beijing, China

References

[1] Koppelaar RHEM, Weikard HP. Assessing phosphate rock depletion and phosphorus recycling options. Global Environmental Change. 2013;**23**(6):1454-1466

[2] Cooper J, Lombardi R, Boardman D, Carliell-Marquett C. The future distribution and production of global phosphate rock reserves. Resources, Conservation and Recycling. 2011;**57**:78-86

[3] van Vuuren DP, Bouwman AF, Beusen AHW. Phosphorus demand for the 1970-2100 period: A scenario analysis of resource depletion. Global Environmental Change. 2010;**20**:428-439

[4] Sverdrup HU, Ragnarsdottir KV. Challenging the planetary boundaries. II: Assessing the sustainable global population and phosphate supply, using a system dynamics assessment model. Applied Geochemistry. 2011;**26**(Supplement):307-310

[5] Chowdhury RB, Moore GA, Weatherley AJ, et al. Key sustainability challenges for the global phosphorus resource, their implications for global food security, and options for mitigation. Journal of Cleaner Production. 2017;**140**:945-963

[6] Kaushik V, Mcnamara PJ, Mayer BK. Meta-analysis of non-reactive phosphorus in water, wastewater, and sludge, and strategies to convert it for enhanced phosphorus removal and recovery. Science of the Total Environment. 2018;**644**:661-674

[7] Jeong YS, Matsubae-Yokoyama K, Kubo H, et al. Substance flow analysis of phosphorus and manganese correlated with South Korean steel industry. Resources, Conservation and Recycling. 2009;**53**(9):479-489

[8] Cordell D, White S. Life's bottleneck: Sustaining the World's phosphorus for a food secure future. Annual Review of Environment and Resources. 2014;**39**(1):161-188

[9] Melia PM, Cundy AB, Sohi SP, et al. Trends in the recovery of phosphorus in bioavailable forms from wastewater. Chemosphere. 2017;**186**:381-395

[10] Roy ED. Phosphorus recovery and recycling with ecological engineering: A review. Ecological Engineering. 2017;**98**:213-227

[11] Cornel P, Schaum C. Phosphorus recovery from wastewater: Needs, technologies and costs. Water Science and Technology. 2009;**59**(6):1069

[12] Desmidt E, Ghyselbrecht K, Zhang Y, et al. Global phosphorus scarcity and full-scale P-recovery techniques: A review. Critical Reviews in Environmental Science and Technology. 2015;**45**(4):336-384

Section 2

Microscopic Study of Phosphorus

Chapter 2

Phosphorus Microbial Solubilization as a Key for Phosphorus Recycling in Agriculture

Agnieszka Saeid

Additional information is available at the end of the chapter

http://dx.doi.org/10.5772/intechopen.81487

"We may be able to substitute nuclear power for coal, and plastics for wood, and yeast for meat, and friendliness for isolation—but for phosphorus, there is neither substitute nor replacement".

Isaac Asimov, The Relativity of Wrong, 1975

Abstract

The chapter describes the phenomenon of microbial solubilization carried out both by bacteria and fungi in the natural environment. The mechanism of dissolving phosphorus compounds by microorganisms, the importance of phosphorus solubilization for its accessibility to plants, as well as the possibility of using microbial solubilization in the process of valorization of secondary raw materials were discussed in this chapter. It was underlined that by using the biofertilizers, we can be able to reduce the amount of artificial fertilizer needed for cropping. Moreover, few models were mentioned to describe this phenomenon to express the changes observed during the solubilization process.

Keywords: microbial solubilization, phosphorus fertilizers, biofertilizers, soil microorganism, *Bacillus*

1. Introduction

The need for mitigation of dependence on phosphate rock of fertilizer industry is one of the main issues to be solved in the next years. Current trends in agriculture are focused on enhancing the efficiency of fertilizer use since approximately 50% of applied mineral fertilizers are lost from the plant-soil system through gaseous emissions, runoff, erosion, and leaching [1]. Green revolution aims to enhance crop yields, improving soil fertility through better management practices, breeding crops with greater tolerance to edaphic stresses and by the development of new inputs based on optimization of the biological/microbiological process [2].

Fertilizer	Content of phosphorus (as P_2O_5) %, w/w	pH	Retrogradation
Single superphosphate (SSP)	18	Neutral	Fast
Granulated superphosphate	19	Neutral	Fast
Superphosphate "SuperMag"	14	Neutral	Fast
Triple superphosphate (TSP)	46	Neutral	Slow
Triple superphosphate granulated, boronated	44	Neutral	Slow
Phosphate rock flours	10–30	Neutral	Slow
Ammonium polyphosphate (APP)	34	Slightly acidic	Fast
Diammonium phosphate (DAP)	46	Slightly acidic	Fast

Table 1. Characteristic of standard phosphorus fertilizers.

Phosphorus is one of the main microelements in plants nutrition, applied to the soils in the form of phosphorus fertilizers that are mainly produced from phosphate rock. In **Table 1**, the most popular form/types of phosphorus fertilizers used nowadays were presented.

Many initiations were raised during past decade to increase the awareness of need for developing the recycling of phosphorus in the industrial scale. The European Union in 2014 (EU) has included phosphorite within the list of 20 critical raw materials [3] since the EU countries are largely dependent on their import. Moreover, recycling of phosphorus from sewage sludge and slaughter waste has been mandatory in Switzerland, Germany, and Austria. Since Switzerland banned the direct application of sewage sludge onto the soil in 2006, its regulation will result in technical recovery and recycling in the form of inorganic products [4].

There is a growing need for ecological engineering approaches that go beyond phosphorus retention to create pathways for phosphorus recovery and recycling, supporting both eutrophication control and food recovery and recycling with ecological engineering [5].

Phosphate rock plays the main source of phosphorus in the chemical industry and its resources are finite and only five countries mainly control its nonrenewable reserves: Morocco, China, Algeria Syria, and Jordan. Since Europe is highly dependent on phosphate rock import, PR was listed within the critical raw materials as it is assessed against its economic importance and its supply risk. At the first stage, phosphate rock is used to produce phosphoric acid, either by treatment with a furnace or with a sulfuric, phosphoric, or nitric acid to produce phosphoric acid. Obtained phosphoric acid is then used in the production of different kinds of fertilizer mainly by the reaction of phosphate rock with phosphoric acid. Since the phosphate rock is critical "spot" in fertilizer industry, many efforts have been done to find alternative sources of phosphorus or/and more sustainable methods of its production by the utilization of more environmentally friendly approach [6].

According to many specialists, its deposits are scarce and according to the most optimistic forecast, geographically, economically, and logistically available resources would be finished within the 150 years. At the same time, there are large amounts of phosphate available in waste streams from, for example, agriculture, sewage treatment, and from industrial side streams, which pose an increasing problem. To avoid wasting these in, for example, landfills, and to counteract the depletion of natural phosphate sources, routes for reuse are explored [7]. In few European countries, the recycling of phosphorus starts to be obligatory (e.g., Austria and Germany). According to the document published on December 17, 2012, the "European Commission has no choice but to go for the transition to a resource-efficient and ultimately regenerative circular economy."

Although the phosphate content of waste streams is usually lower than the typical P_2O_5 content of phosphate rock which is 30–40% (=13–17.5% P), it is necessary to start to apply the circular economy within the phosphorus fertilizers industry since the agriculture is the main recipient of phosphorus [7].

Phosphorus in the soil is present in organic forms and consists 20–80% of the total. Often, less than 10 μM of phosphorus in soil solution is available to plants, while the mineral fraction of phosphorus from soil can be available to plant, but via the desorption and solubilization process [8]. Integrated soil fertility management (ISFM) is a soil management approach that emphasizes the combined application of organic and mineral fertilizer inputs with the goal of improving yields and fertilizer use efficiency. Both organic and inorganic fertilizers were and are widely applied to sustain soil productivity [9].

2. Microbial solubilization

Unlike inorganic fertilizers, organic nutrients from the plant and animal-based residues are often not readily available to the plant and must be converted into plant available forms by microbes in the substrate.

One of the methods that could be efficiently used as a way to mitigate the phosphorus problem is the utilization of natural microorganisms that in the native environment are able to increase phosphorus availability for its cells as well as for root plants, which is called microbial solubilization. The natural ability of soil microorganism to produce organic acids, enzymes that released into the soil environment, results in increasing the availability of many nutrients that in the soil are present in the retrogradative form such as phosphorus, as well as many micronutrients such as zinc or selenium [10].

Plants can uptake the phosphorus or any other nutrients from the soil when is soluble/present in the soil solution. Phosphorus fertilizer used as a soil conditioner delivered phosphorus in available-to-plant form but dissolved in the soil solution and in the presence of Al(III) and Fe(II), it forms unavailable-to-plant compounds and becomes retrogradative [11]. According to many specialists, the amount of phosphorus retrograde in the soil is so abundant that there is no need to use fertilizers. The solution to cover the requirements of plants would be to "activate" retrogradative form of phosphorus into available, for example, by utilization of soil microorganism. Unavailable phosphorus present in the soil in organic as well as inorganic form is launched via solubilization of inorganic form and mineralization of organic forms of phosphorus [12]. Thus, the residual phosphate fertilizers in the soil can be well utilized and the external application should be optimized. Microbial solubilization is performed by the production of acids that by decreasing pH and attraction of cations from structure that effects with the release of phosphorus [13].

Many products are nowadays available on the market that delivers the soil microorganisms in concentrated forms such as lyophilisate or single bacterial strain or consortium. Such beneficial microorganism is so-called biofertilizers, which are classified as biostimulants. According to the definition proposed by Vessey [14], biofertilizers are substances which contain living microorganisms which, when applied to seed, plant surfaces, or soil, colonize the rhizosphere or the interior of the plant, and promote growth by increasing the supply or availability of primary nutrients to the host plant. Biofertilizers, regardless of its form, can be applied in three ways (**Figure 1**) that are related to the stage of growth of the plant to be inoculated. Biofertilizers are produced in liquid, powder, and granular forms and applied to soil, compost, seed, seedling, and plant leaves.

Many bioproducts which activity is based on soil microorganism are available on the market (**Table 2**). According to the industrystock.com database, 35 companies were listed as the ones that are related to the production of biofertilizers.

"Microbial consortia," are in other words biopreparations of various microorganisms, are a beneficial microbial mixture that supports plant nutrition and health, becoming more popular. Especially when combined with organic materials, they can serve as organic fertilizers. Some of the effects of combinations of bioorganic fertilizers were discussed elsewhere [15]. The application of biofertilizers has illustrated that 50% of recommended NPK fertilizers can be reduced through inoculation with microbial consortia with no adverse effects on growth, nutrition, or yield [16]. The mechanisms by which these preparations enhance plant growth are numerous; the main mechanism of solubilization of inorganic phosphorus compounds is the activity of organic acids, siderophores, protons, hydroxyl ions, and CO_2 [17].

In this context, the reduced use of chemical fertilizers with increased application of organic fertilizers is considered a compulsory route to alleviate the pressure on the environment derived from agricultural practice [18]. Such an approach could mitigate the problem of depletion of phosphate rock. Microbial solubilization could be also utilized in the "activation"

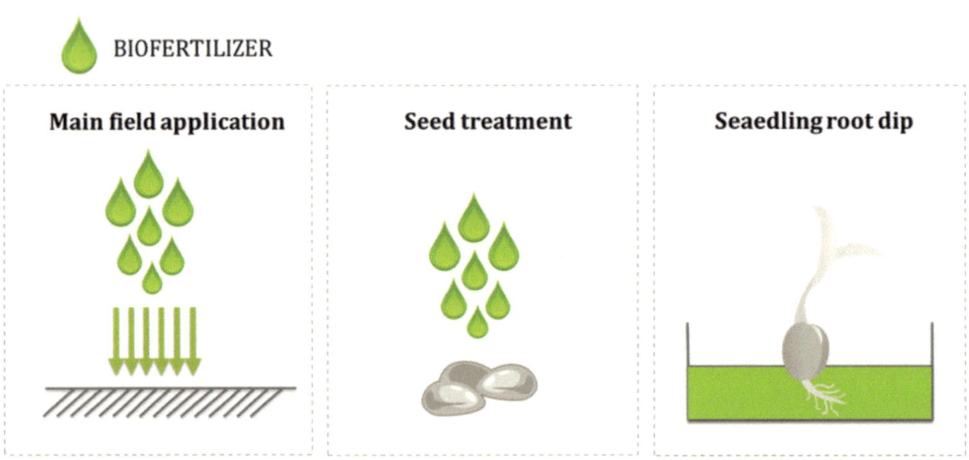

Figure 1. Application of biofertilizers.

No.	Product	Producer	Content		Effect
1	Bactim receptor	Intermag, Poland	*Glomus coronatum, Glomus caledonium, Glomus mosseae, Glomus viscosum, Rhizophagus irregularis, Streptomyces* spp., *Streptomyces avernichilis, Pochonia chlamydosporia*	7.2×10^6 CFU/g	• It limits damage to the root system through nematodes. • Increases plant resistance to root system diseases. • Limits the accumulation of harmful substances in plants.
2	Bactim vigor	Intermag, Poland	*Glomus coronatum, Glomus caledonium, Glomus mosseae, Glomus viscosum, Rhizophagus irregularis, Bacillus subtilis, Streptomyces spp., Trichoderma harzianum, Trichoderma viride, Pichia pastoris*	10.2×10^7 CFU/g	• Increases the resistance of field vegetables and fruit trees and bushes to abiotic and biotic stresses. • It improves soil structure and increases the availability of nutrients from the soil. • It improves the quality and postharvest life of crops.
3	Bactim starter	Intermag, Poland	*Azospirillum brasilense, Azotobacter chroococcum, Bacillus subtilis, Bacillus megaterium, Pseudomonas fluorescens*	2.5×10^{10} CFU/g	• Supports seed germination, ensures faster and better plant emergence. • Improves the supply of nitrogen seedlings. • Provides better phosphorus uptake by the young root system.
4	Acetobacter spp. fertilizers	Jay Ambe Agro Products, India	*Gluconacetobacter diazotrophicus*	1×10^9 CFU/ml	• Produces growth promoting substances such as indole acetic acid (IAA) and gibberellins that promote root proliferation and increase the rootlet density and root branching which resulting in increased uptake of mineral and water which promotes cane growth and sugar recovery from the cane.
5	Azospirillum fertilizer	Jay Ambe Agro Products, India	*Azospirillum brasilense*	1×10^9 CFU/ml	• Actively fix atmospheric nitrogen through asymbiotic relation with the leguminous plants.

Table 2. Examples of bioproducts available on the market.

of phosphorus bound in the by-products originated from agriculture, food processing, or from ashes obtained from incineration of different types of wastes that are characterized by high content of phosphorus.

Several technologies for solubilization of phosphorus compounds have been developed that can be described as processes which transform phosphorus into bioavailable forms via a range of chemical reactions and biological interactions [19]. They include phosphate-solubilizing microorganisms, phosphatase enzymes and enzyme activators, low-molecular-weight organic acids, humic acids, lignin, crop residues, biochar, and zeolites. The goal is to elaborate integrated procedure/P recovery techniques to treat the waste stream to generate a product more suitable as a soil amendment [20].

According to the data evaluated by Koppelaar and Weikard [21], applying phosphorus-solubilizing biofertilizing microorganisms results in 10% reduction of phosphorus fertilizer application. Zabihi et al. [22] demonstrated the possibility to reduce up to 50% of fertilizer application without crop yield reduction under controlled conditions. Unfortunately, results are contrasting between greenhouse and field experiments possibly due to plant-soil interaction knowledge gaps, differences between plant and soil types, and incorrect microorganism strain selection [23].

The efficiency of microbial solubilization can be enhanced by the presence of numerous compounds. Humic substances that comprise a major part of organic matter and their influence on soil properties are well known and could be used to improve microbial activity [24]. Phosphate-solubilizing microorganisms, selected for RP solubilization, combined with humic acid, positively stimulated root and shoot weight compared with noninoculated plants by 17 and 22%, respectively. Despite this biomass increase, no difference was observed in P concentration, indicating an increased P use efficiency. The application of both PSM and HS with RP may be a suitable method for reduction of soluble P fertilizer demands without compromising plant yields [25]. Results described in the literature show a promising use of humic substances to improve the benefit of phosphorus solubilizing microorganism [26]. Zeolites seem to be also one of the materials that have the potential to improve the results of solubilization since it was used as an activator of phosphate rock. The zeolite/rock phosphate combination possibly acted as an exchange fertilizer, with Ca exchanging onto the zeolite in response to plant uptake of nutrient cations enhancing the dissolution of the rock phosphate [27, 28]. Another modification of the solubilization process is its combination with composting what can provide many potential agronomic benefits (e.g., slow-release nutrients) [29].

3. Phosphorus by-products

Waste streams originating from the food industry (e.g., agricultural runoff, stormwater, animal manures, food and food-processing wastes, human urine and feces, municipal wastewater, biosolids) [5] seem to one of the most interesting sources of phosphorus that could be valorized into valuable products such as fertilizers. Phosphorus in the wastes is present in unavailable-to-plants form and needs to be treated before applied. The methods that are described and successful consider the solubilization of phosphorus compounds present in the phosphate rock [30]. The advantage of this approach is the production of phosphorus fertilizer without the problematic side products such as phosphogypsum but it is not solving the problem with the depletion of phosphate rock which is still the most important raw material for phosphorus fertilizer industry. Another solution is the utilization of solubilization of secondary raw materials such as meat-bone meal [31], ashes of the waste sludge from wastewater treatment plant with the secondary stage of biological treatment process [32], bones and fish bones [33].

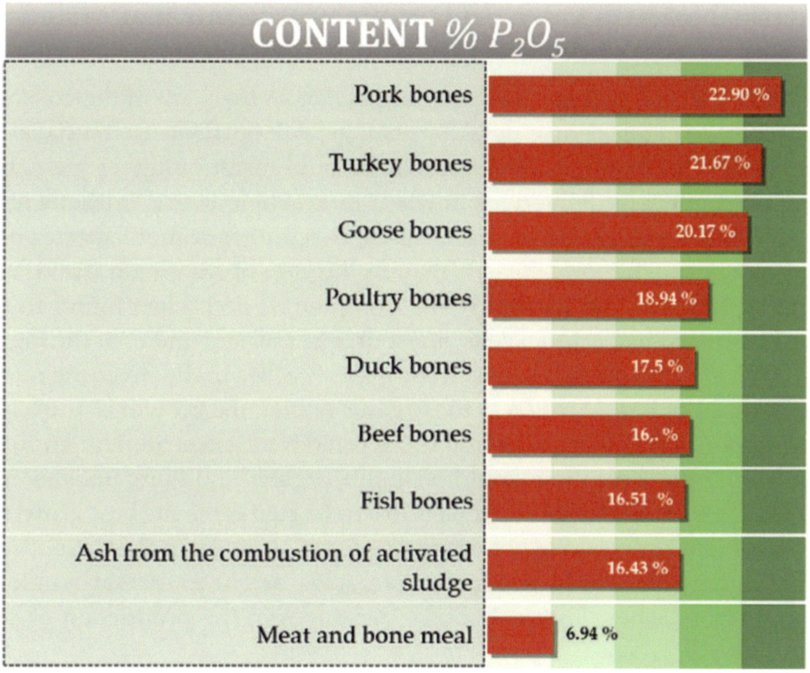

Figure 2. The content of P_2O_5 in P-bearing renewable raw materials.

Phosphorus from phosphate rock that has been utilized in the agriculture and fertilizer or chemical industry is present in the following forms:

- retrogradative form present in soil as a result of not proper fertilization;
- wastes from food production, such as bones and fish bones;
- ashes originated from incineration of activated sludge from wastewater treatment plant or wastes originated from the slaughterhouse.

Mentioned streams of phosphorus seem to be the most concentrated ones, thus have been chosen to be the secondary source of phosphorus in the circular economy strategy [7, 34].

Figure 2 presents the content of phosphorus expressed as P_2O_5 in few phosphorus renewable raw materials; its exact content can significantly vary and is strongly related to the origin, kind of animals, its nutritional status, diet, etc. In **Figure 2**, last bar—meat-bone meal—was added to compare the P_2O_5 content and to prove that it is better to use the original raw materials without treatment since it causes significant phosphorus losses.

The second sector, after the agriculture, that uses phosphorus obtained from phosphate rock is private households. It generates a significant amount of wastewaters; after the treatment in a wastewater treatment plant, almost all phosphorus is removed by chemical and biological treatment methods and is fixed in the sewage sludge. Since its direct application in agriculture is no longer accepted, incineration is undertaken that generated significant amounts of ashes with up to 21% P_2O_5 in the inorganic form [5].

4. Phosphorus biofertilizers based on renewable raw materials

Introduction of beneficial and natural microorganisms in the form of the consortium with confirmed properties of releasing crucial for plant growth nutrients from unavailable forms seems to be the best solution. Such an approach would result in higher yield, better quality of plants as an effect of delivery of nutrient in available forms, without utilization of phosphate rock mining products. In many cases, two, three, or more species of beneficial microorganism acts much more efficient than in the case of the single strain. Some of the sources of phosphorus are rich with organic compounds and when found in the growth medium can be utilized by bacterial cells as a source of nutrient and then the higher growth is observed, such as bones or fish bones. In the case of ashes that deliver the nutrients only in inorganic form since it is deprived of the organic matter, the growth is much lower when compared with bones and fish bones but at the same time ashes are known to have more than 20% of P_2O_5. A good solution would be to mix organic and inorganic raw materials to ensure at the same time source of valuable nutrient to guarantee the best growth condition and deliver the high dose of phosphorus in the form of ashes. In this case, the enhanced growth of bacterial cells would produce more acids or enzymes that would solubilize the phosphorus from ashes. **Figure 3** shows the stages in the production of phosphorus biofertilizers [35–48].

Nine formulations were obtained based on the single bacterial strain, such as *B. megaterium* or *A. ferrooxidans* as shown in **Figure 2**. Three different raw materials were used as a source of

Figure 3. The general scheme of stages in the solubilization.

Fertilizer	Form	Raw material	Bacteria	P_2O_5, % mass
1	Suspension	Ash	*B. megaterium*	0.406
2	Suspension	Bones	*B. megaterium*	0.595
3	Suspension	Ash	*B. megaterium*	0.406
4	Granular	Ash	*A. ferrooxidans*	21.2
5	Granular	Ash + bones	*A. ferrooxidans*	17.2
6	Granular	Ash + bones	*B. megaterium*	13.5
7	Granular	Ash + blood	*B. megaterium*	21.9
8	Granular	Ash + blood	*B. megaterium*	11.3
9	Granular	Ash + blood	*B. megaterium*	11.3

Table 3. The content of P_2O_5 in biofertilizer formulation.

phosphorus: bones and ash but also blood meal, which was used as a source of nitrogen but at the same time played a role of binder in the granular formulations.

The content/concentration of P_2O_5 in obtained phosphorus biofertilizers based on the *Acidithiobacillus ferrooxidans* and *Bacillus megaterium* was presented in **Table 3**.

Obtained formulations in semitechnical scale were tested in the field test. More details concerning the production process [49] as well as the more detailed results from field tests [50] were described elsewhere.

In terms of the impact on the yield of test plants, phosphorus biofertilizers from renewable raw materials were comparable to commercial fertilizers: moreover, they did not affect adversely on the morphological and physiological features of the test plants; did not affect the degree of weed infestation, infection by fungal pathogens, and lodging of test plants; also, did not affect the humidity, temperature, salinity, and pH of the soil, the total number of heterotrophic bacteria and fungi in the soil, and the presence of earthworms; and did not change the elemental composition of plants and soil. Utilization of biofertilizers based on the *B. megaterium* did not increase the number of *Bacillus megaterium* in the soil, but they were conducive to the stabilization of this strain in the soil environment that is crucial since any pressure affected on the homeostasis of soil environment has an adverse effect [50].

5. Modeling of solubilization

The process of microbial solubilization can be defined as the transformation of insoluble compounds present in inorganic form to a soluble form, and thus available to plants. As a result of the activity of microorganisms, dissolved minerals are formed. The production of organic acids results in acidification of the microbial cell and its surroundings by decreasing the pH. The amount of soluble phosphate released depends on the strength and type of produced acids [51].

Parameters that express and characterize the solubilization are used to evaluate the effectiveness and are as follows: pH, μ (day^{-1}), and CP_2O_5. To describe experimental data, a few models were evaluated to understand the changes in the process [43–45].

Acids produced by bacteria caused an increase in the concentration of hydrogen ions and higher release of P_2O_5. Decreasing pH as a result of production of acids was described in Eq. (1):

$$pH = f(C_{P_2O_5}) = \frac{A + pH_{min} \cdot C_{P_2O_5}}{C_{P_2O_5}} \tag{1}$$

where A (mg/L) is a constant describing the decay of curve. Evaluated value of pH_{min} can be interpreted as the minimal value of pH. The P_2O_5 concentration is correlated with the pH of the liquid phase through the following model (**Figure 4**).

To describe the changes in P_2O_5 concentrations during solubilization, the proposed model that describes kinetics of releasing phosphorus (expressed as the P_2O_5) was used (Eq. (2)):

$$C_{P_2O_5} = f(t) = \frac{C_{P_2O_5}^{max}}{1 + b \cdot exp^{-k \cdot t}} \tag{2}$$

where the $C_{P_2O_5}^{max}$ (mg/L) is the maximum concentration of P_2O_5, b is the constant that expresses time when $C_{P_2O_5}$ is equal to ½ of $C_{P_2O_5}^{max}$, and k (1/day) constant is the variable slope, which is called the Hill slope. When k is greater, the curve changes more sharply and it means that solubilization process proceeds faster (**Figure 5**).

In case the solubilization is conducted in the columns in the *in situ* studies, the concentration of released phosphorus expressed as the amount of P_2O_5 is monitored versus the amount of remaining in the column that was strongly correlated with the pH of the liquid phase through Eq. (3), (**Figure 6**):

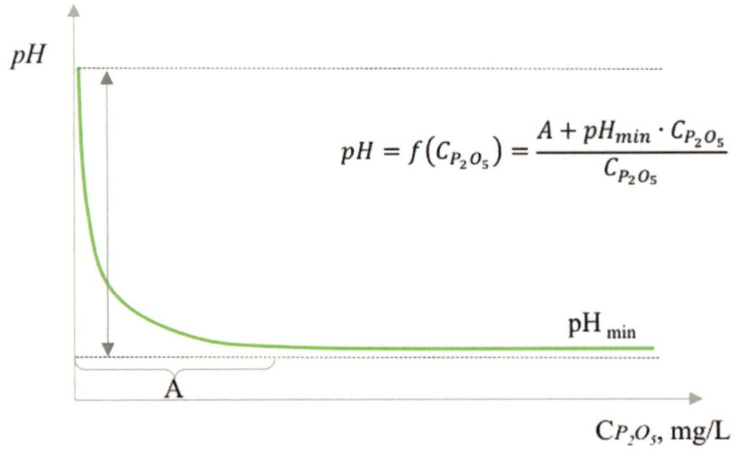

Figure 4. Description/visualization of proposed model $f(C_{P_2O_5}) = pH$.

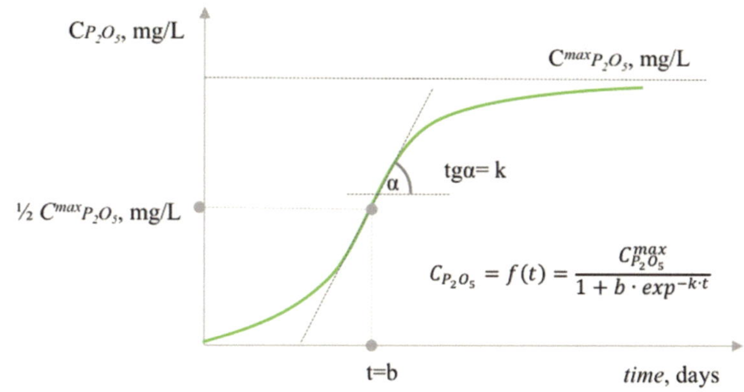

Figure 5. Description/visualization of proposed model $f(t) = C_{P_2O_5}$.

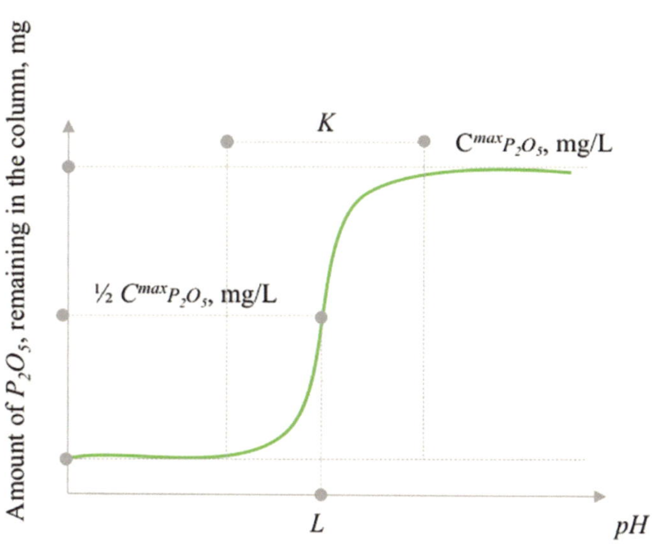

Figure 6. Description/visualization of proposed model $f(pH) = C_{P_2O_5}$.

$$C_{P_2O_5} = C^{min}_{P_2O_5} \frac{C^{max}_{P_2O_5} - C^{min}_{P_2O_5}}{1 + e^{-\left(\frac{pH-L}{K}\right)}} \quad (3)$$

This function has four parameters. $C^{min}_{P_2O_5}$ and $C^{max}_{P_2O_5}$ are two asymptotic values. The curve crosses over between these two asymptotic values in a region of pH whose approximate width is K and which is centered around L. $C^{min}_{P_2O_5}$ and $C^{max}_{P_2O_5}$ (mg/L) are constants describing the decay of curve. Evaluated value of L can be interpreted as value of pH that corresponds with the half of phosphorus (expressed as P_2O_5) that under considered condition could be released/solubilized [43].

As the solubilization progresses, the unavailable form of phosphorus becomes more available to plants and the same for bacterial cells that are responsible for phosphorus solubilization. That is why, measuring the concentration of P_2O_5 only in the solution as an evaluation of

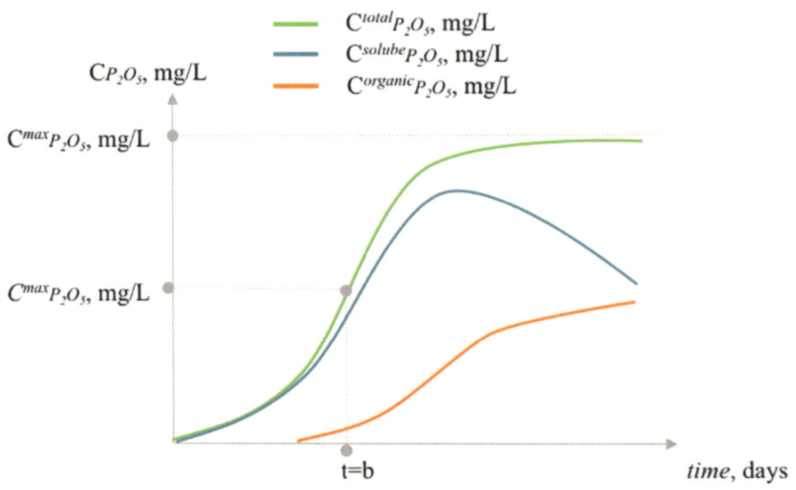

Figure 7. Kinetics of forming/liberating different forms of phosphorus during the microbial solubilization of inorganic forms of phosphorus.

solubilization effective is not so much accurate since some part of liberated phosphorus from unavailable forms was used by bacterial cells to maintain its own metabolism since phosphorus is one of the main macroelements not only for the plant. That is why, the following fraction of phosphorus should be distinguished according to **Figure 7**.

Phosphorus present in the biomass of bacterial cells in the organic form as a result of its immobilization can become available to plants in the mineralization process. To evaluate the efficiency of microbial solubilization, both organic and soluble forms of phosphorus should be concerned and included in the empirical model, for better understanding this phenomenon.

6. Conclusions

The natural ability of soil microbiota to solubilize nutrients present in the soil is the crucial link in the nutrient cycling in the natural environment. By the incorporation of this phenomenon with the processing of P-bearing renewable raw materials, it is possible to produce a source of available forms of phosphorus that could be applied in agriculture as a soil conditioner. Utilization of one of the components of the subsystem of saprotrophs to accelerate the circulation of phosphorus seems to be the easiest and the most beneficial for the natural environment. Since the by-products that carry a significant load of phosphorus are abundant and are produced in significant amount, every year, all over the world, they should be utilized or valorized into valuable products; such an attitude could serve as an efficient way of introduction of one of the most important nutrients in plant growth into the soil in available form, that could significantly limit the number of chemical fertilizers that should be used to cover the nutritional requirements of plants. Moreover, this proposal is within the circular economy approach that nowadays is in the main direction of developing new methods of treatment of by-products to lessen the amount of raw materials used in the production process.

Acknowledgements

This work was financed within the framework of Grant PBS 2/A1/11/2013 entitled: "Phosphorus renewable raw materials—a resource base for the new generation of fertilizers" awarded by the National Centre for Research and Development. This project was supported by the Wroclaw Centre of Biotechnology program. The Leading National Research Centre (KNOW) from 2014 to 2018. The following coworkers should be acknowledged for their cooperation: Magdalena Jastrzębska, Kinga Treder, Przemysław Makowski, Wiesław P. Jastrzębski, Małgorzata Wyciszkiewicz, and Sylwia Baśladyńska.

Author details

Agnieszka Saeid

Address all correspondence to: agnieszka.saeid@pwr.edu.pl

Faculty of Chemistry, Department of Advanced Material Technologies, Wroclaw University of Science and Technology, Wroclaw, Poland

References

[1] Agriculture and Environment - Pollution Risks - Statistics Explained. Available from: http://ec.europa.eu/eurostat/statistics-explained/index.php/Agriculture_and_environment_-_pollution_risks [Accessed: 01-08-2019]

[2] Canellas LP, Olivares FL. Physiological responses to humic substances as plant growth promoter. Chemical and Biological Technologies in Agriculture. 2014;**1**(1):3

[3] Report on Critical Raw Materials for the EU. Report of the Ad hoc Working Group on Defining Critical Raw Materials. European Commission; 2014

[4] European Sustainable Phosphorus Platform, Scope and News. Switzerland Makes Phosphorus Recycling Obligatory; 2015

[5] Roy ED. Phosphorus recovery and recycling with ecological engineering: A review. Ecological Engineering. 2017;**98**:213-227. DOI: 10.1016/j.ecoleng.2016.10.076

[6] Schipper WJ, Klapwijk A, Potjer B, Rulkens WH, Temmink BG, Kiestra FDG, et al. Phosphate recycling in the phosphorus industry. Phosphorus Research Bulletin. 2004; **15**:47-51. DOI: 10.3363/prb1992.15.0_47

[7] Rechberger H. Practical Handbook of Material Flow Analysis. Florida, USA: CRC Press LLC; 2004

[8] Shen J, Yuan L, Zhang J, Li H, Bai Z, Chen X, et al. Phosphorus dynamics: From soil to plant. Plant Physiology. 2011;**156**(3):997-1005. DOI: 10.1104/pp.111.175232

[9] Hobbs PR, Sayre K, Gupta R. The role of conservation agriculture in sustainable agriculture. Philosophical Transactions of the Royal Society B. 2008;**363**:543-555. DOI: 10.1098/rstb.2007.2169

[10] Malusá E, Sas-Paszt L, Ciesielska J. Technologies for beneficial microorganisms inocula used as biofertilizers. The Scientific World Journal. 2012. Article ID: 491206. DOI: 10.1100/2012/491206. [Accessed: 01-08-2019]

[11] Balemi T, Negisho K. Management of soil phosphorus and plant adaptation mechanisms to phosphorus stress for sustainable crop production: A review. Journal of Soil Science and Plant Nutrition. 2012;**12**(3):547-562. DOI: 10.4067/S0718-95162012005000015

[12] Menzies N. The Science of Phosphorus Nutrition: Forms in the Soil, Plant Uptake, and Plant Response. Available from: https://grdc.com.au/resources-and-publications/grdc-update-papers/tab-content/grdc-update-papers/2009/02/the-science-of-phosphorus-nutrition-forms-in-the-soil-plant-uptake-and-plant-response [Accessed: 01-08-2019]

[13] Sharma SB, Sayyed RZ, Trivedi MH, Gobi TA. Phosphate solubilizing microbes: Sustainable approach for managing phosphorus deficiency in agricultural soils. Springerplus. 2013;**2**:587. DOI: 10.1186/2193-1801-2-587

[14] Vessey JK. Plant growth promoting rhizobacteria as biofertilizers. Plant and Soil. 2003; **255**:571-586. DOI: 10.1023/A:1026037216893

[15] Wallenstein MD, Bell CW. Synergistic Bacterial Consortia for Mobilizing Soil Phosphorus. US Application, US20160145163A1; 2014

[16] Duarah I, Deka M, Saikia N, Boruah HPD. Phosphate solubilizers enhance NPK fertilizer use efficiency in rice and legume cultivation. Biotechnology. 2011;**1**(4):227-238. DOI: 10.1007/s13205-011-0028-2

[17] Alori ET, Glick BR, Babalola OO. Microbial phosphorus solubilization and its potential for use in sustainable agriculture. Frontiers in Microbiology. 2017;**8**:971. DOI: 10.3389/fmicb.2017.00971

[18] Martínez-Alcántara B, Martínez-Cuenca M-R, Bermejo A, Legaz F, Quiñones A. Liquid organic fertilizers for sustainable agriculture: Nutrient uptake of organic versus mineral fertilizers in citrus trees. PLoS One. 2016;**11**(10):e0161619. DOI: 10.1371/journal.pone.0161619

[19] Zhu J, Li M, Whelan M. Phosphorus activators contribute to legacy phosphorus availability in agricultural soils: A review. Science of the Total Environment. 2018;**612**: 522-537. DOI: 10.1016/j.scitotenv.2017.08.095

[20] Wei Y, Zhao Y, Shi M, Cao Z, Lu Q, Yang T, et al. Effect of organic acids production and bacterial community on the possible mechanism of phosphorus solubilization during composting with enriched phosphate-solubilizing bacteria inoculation. Bioresource Technolnology. 2018;**247**:190-199. DOI: 10.1016/j.biortech.2017.09.092. 2017

[21] Koppelaar RHEM, Weikard HP. Assessing phosphate rock depletion and phosphorus recycling options. Global Environmental Change. 2013;**23**(6):1454-1466. DOI: 10.1016/j.gloenvcha.2013.09.002

[22] Zabihi HR, Savaghebi GR, Khavazi K, Ganjali A, Miransari M. Pseudomonas bacteria and phosphorous fertilization, affecting wheat (*Triticum aestivum L.*) yield and P uptake under greenhouseand field conditions. Acta Physiologiae Plantarum. 2011;**33**:145-152. DOI: 10.1007/s11738-010-0531-9

[23] Uribe D, Sánchez-Nieves J, Vanegas J. Role of microbial biofertilizers in the development of a sustainable agriculture in the tropics. In: Pion D, editor. Soil Biology and Agriculture in the Tropics. Berlin: Springer-Verlag; 2010

[24] Canellas LP, Balmori DM, Médici LO, Aguiar NO, Campostrini E, Rosa RCC, et al. A combination of humic substances and *Herbaspirillum seropedicae* inoculation enhances the growth of maize (*Zea mays* L.). Plant and Soil. 2013;**366**(1-2, 132):119. DOI: 10.1007/s11104-012-1382-5

[25] Giro VB, Jindo K, Vittorazi C, Canellas LP, Olivares FL. Rock phosphate combined with phosphate solubilizing microorganisms and humic substance for reduction of plant phosphorus demands from single superphosphate. Acta Horticulturae. 2016;**1146**:63-68

[26] Olivares FL, Busato JG, de Paula AM, Lima LS, Aguiar NO, Canellas LP. Plant growth promoting bacteria and humic substances: Crop promotion and mechanisms of action. Chemical and Biological Technologies in Agriculture. 2017;**4**:30. DOI: 10.1186/s40538-017-0112-x. [Accessed: 01-08-2019]

[27] Ramesh K, Reddy DD. Zeolites and their potential uses in agriculture. Advances in Agronomy. 2011;**113**:219-241. DOI: 10.1016/B978-0-12-386473-4.00004-X

[28] Yang M, Lin J, Zhan Y, Zhu Z, Zhang H. Immobilization of phosphorus from water and sediment using zirconium-modified zeolites. Environmental Science and Pollution Research. 2015;**22**:3606. DOI: 10.1007/s11356-014-3604-2

[29] Wyciszkiewicz M, Saeid A, Samoraj M, Chojnacka K. Solid-state solubilization of bones by *B. megaterium* in spent mushroom substrate as a medium for a phosphate enriched substrate. Journal of Chemical Technology and Biotechnology. 2017;**92**:1397-1405. DOI: 10.1002/jctb.5135

[30] Shen J, Yuan L, Zhang J, Li H, Bai Z, Chen X, et al. Phosphorus dynamics: From soil to plant, focus issue on phosphorus plant physiology. Plant Physiology. 2011;**156**(3): 997-1005. DOI: 10.1104/pp.111.175232

[31] Kimberly AK, Henry GT. Comparison of bone products for phosphorus availability. HortTechnology. 1996;**6**(3):257-260. Available from: https://pdfs.semanticscholar.org/d7d5/31f7a3bf73d4fef921e83a4a8e44b5a4334d.pdf [Accessed: 01-08-2019]

[32] Stark K. Phosphorus Release from Sewage Sludge by Use of Acids and Bases. KTH Royal Institute of Technology. Available from: https://www.kth.se/polopoly_fs/1.650616!/JPS9p19.pdf [Accessed: 01-08-2019]

[33] Saeid A, Labuda M, Chojnacka K. Valorization of bones to liquid phosphorus fertilizer by microbial solubilization. Waste and Biomass Valorization. 2014;**5**(2):265-272. DOI: 10.1007/s12649-013-9238-7

[34] Karunanithi R, Szogi AA, Bolan N, Naidu R, Krishnamoorthy S. Phosphorus recovery and reuse from waste streams. Advances in Agronomy. 2015;**131**:173-250. DOI: 10.1016/bs.agron.2014.12.005

[35] Labuda M, Saeid A, Chojnacka K, Górecki H. Zastosowanie *Bacillus megaterium*, w solubilizacji fosforu. Przemysł Chemiczny. 2012;**91**(5):837-840

[36] Saeid A, Chojnacka K, Labuda M, Górecki H. Procesy biotechnologiczne w wytwarzaniu nawozów fosforowych. Przemysł Chemiczny. 2012;**91**(5):952-955

[37] Saeid A, Chojnacka K, Labuda M, Górecki H. Wykorzystanie mikroorganizmów w wytwarzaniu nawozów fosforowych. Przemysł Chemiczny. 2012;**91**(5):956-958

[38] Saeid A, Chojnacka K, Labuda M, Górecki H. Ocena właściwości użytkowych nowego bionawozu fosforowego. Przemysł Chemiczny. 2013;**92**(7):1311-1314

[39] Wyciszkiewicz M, Saeid A, Chojnacka K, Górecki H. Ocena skuteczności bionawozu fosforowego w testach kiełkowania. Przemysł Chemiczny. 2014;**93**(7):1210-1214

[40] Saeid A, Jastrzębska M, Wyciszkiewicz M, Chojnacka K, Górecki H. Koncepcja wytwarzania nowej generacji bionawozów fosforowych - projekt BioFertP. Przemysł Chemiczny. 2015;**94**(3):361-365

[41] Wyciszkiewicz M, Saeid A, Chojnacka K, Górecki H. New generation of phosphate fertilizer from bones, produced by bacteria. Open Chemistry. 2015;**13**:951-958. DOI: 10.1515/chem-2015-0113

[42] Wyciszkiewicz M, Saeid A, Chojnacka K, Górecki H. Production of phosphate biofertilizers from bones by phosphate-solubilizing bacteria *Bacillus megaterium*. Open Chemistry. 2015;**13**:1063-1070. DOI: 10.1515/chem-2015-0123

[43] Wyciszkiewicz M, Saeid A, Chojnacka K. In situ solubilization of phosphorus bearing raw materials by *Bacillus megaterium*. Engineering in Life Science. 2017;**17**:749-758. DOI: 10.1002/elsc.201600191

[44] Wyciszkiewicz M, Saeid A, Malinowski P, Chojnacka K. Valorization of phosphorus secondary raw materials by *Acidithiobacillus ferrooxidans*. Molecules. 2017;**22**:473. DOI: 10.3390/molecules22030473

[45] Wyciszkiewicz M, Saeid A, Chojnacka K. Solubilization of renewable phosphorus sources with organic acids produced by *Bacillus megaterium*. Journal of Renewable Materials. Supplement 2017. p. 1-14. DOI: 10.7569/JRM.2017.634132

[46] Jastrzębska M, Saeid A, Kostrzewska MK, Baśladyńska S. New phosphorus biofertilizers from renewable raw materials in the aspect of cadmium and lead contents in soil and plants. Open Chemistry. 2018;**16**:35-49. DOI: 10.1515/chem-2018-0004

[47] Kostrzewska MK, Jastrzębska M, Treder K, Saeid A, Makowski P, Jastrzębski WP. Evaluation of applicability of a Bacillus megaterium-containing ash and blood-based fertilizer in the light of selected morphological and physiological attributes of spring wheat. Przemysł Chemiczny. 2017;**96**(10):2162-2167

[48] Jastrzębska M, Kostrzewska MK, Saeid A, Treder K, Makowski P, Jastrzębski WP, et al. Granulated phosphorus fertilizer made of ash from biomass combustion and bones with addition of Bacillus megaterium in the field assessment. Part 1. Impact on yielding and sanitary condition of winter wheat. Przemysł Chemiczny. 2017;**96**(10):2168-2174

[49] Rusek P, Rolewicz M, Mikos Szymańska M, Cichy B. Technology of production of bio-fertilizers in pilot scale. In: Saeid A, Chojnacka K, editors. Innovative Phosphorus Bio-Fertilizers. USA: Nova Science; 2019. (in print)

[50] Jastrzębska M, Kostrzewska MK. Utilitarian properties of new fertilizer formulations evaluated in field tests. In: Saeid A, Chojnacka K, editors. Innovative Phosphorus Bio-Fertilizers. USA: Nova Science; 2019. (in print)

[51] Wyciszkiewicz M, Saeid A, Dobrowolska-Iwanek J, Chojnacka K. Utilization of micro-organisms in the solubilization of low-quality phosphorus raw material. Ecological Engineering. 2016;**89**:109-113. DOI: 10.1016/j.ecoleng.2016.01.065

Chapter 3

Simulations Suggest Possible Triply Bonded Phosphorus≡E13 Molecules (E13 = B, Al, Ga, In, and Tl)

Jia-Syun Lu, Ming-Chung Yang and Ming-Der Su

Additional information is available at the end of the chapter

http://dx.doi.org/10.5772/intechopen.77055

Abstract

The effect of substitution on the potential energy surfaces of RE13 ≡ PR (E13 = B, Al, Ga, In, Tl; R = F, OH, H, CH3, SiH3, SiMe(Si*t*Bu3)2, Si*i*PrDis2, Tbt, and Ar* is studied using density functional theory (M06-2X/Def2-TZVP, B3PW91/Def2-TZVP and B3LYP/LANL2DZ + dp). The theoretical results demonstrate that all triply bonded RE13 ≡ PR compounds with small substituents are unstable and spontaneously rearrange to other doubly bonded isomers. That is, the smaller groups, such as R = F, OH, H, CH3 and SiH3, neither kinetically nor thermodynamically stabilize the triply bonded RE13 ≡ PR compounds. However, the triply bonded R′E13≡PR′ molecules, possessing bulkier substituents (R′ = SiMe(Si*t*Bu3)2, Si*i*PrDis2, Tbt and Ar*), are found to have a global minimum on the singlet potential energy surface. In particular, the bonding character of the R′E13≡PR′ species is well defined by the valence-electron bonding model (model [II]). That is to say, R′E13≡PR′ molecules that feature groups are regarded as R′-E13⇆P-R′. The theoretical evidence shows that both the electronic and the steric effects of bulkier substituent groups play a prominent role in rendering triply bonded R′E13≡PR′ species synthetically accessible and isolable in a stable form.

Keywords: phosphorus, group 13 elements, triple bond, substituent effects, valence electrons

1. Introduction

Phosphorus is an interesting element, but many chemists have a poor comprehension of its bonding properties. Even though phosphorus and nitrogen belong to the same group in the periodic table, molecular nitrogen is a triply bonded diatomic molecule, but elemental white phosphorus is a tetrahedral compound wherein each atom is connected by three single bonds to the other atoms in the molecule. Phosphorus is usually connected to other elements by a single chemical bond, which has been verified by lot of experimental evidences [1–14]. Also, molecules that feature a phosphorus double bond have been the subject of many experimental and theoretical studies of structure and reactivity [15–27]. However, little is known about the molecules that feature a phosphorus triple bond [28–32]. In particular, whether it is possible to anticipate the stability of the R-E13 ≡ phosphorus-R (E13 = B, Al, Ga, In, and Tl) species based on the effects of substituents, since the R-E13 ≡ phosphorus-R systems are isoelectronic to the R-E14 ≡ E14-R (E14 = C, Si, Ge, Sn, and Pb) compound from the valence electron viewpoints.

This study uses the heavier acetylene analogue, R-E13 ≡ P-R as a model molecule to determine the possibility of generating stable RE13PR species that feature the E13 ≡ P triple bond. In order to understand the effects of substituents on the stability of triply bonded RE13 ≡ PR molecules, both small and bulky groups are chosen in this work. A better understanding of the bonding character and the structure of triply bonded RE13 ≡ PR species will allow experimental chemists to discover novel and stable molecules that feature the E13 ≡ P triple bond.

1.1. General considerations

This section uses a simple valence-electron bonding model to demonstrate the bonding nature of substituted triply bonded RE13≡PR compounds.

First, the RE13≡PR species is separated into two units: R-E13 and R-P. **Figure 1** shows that these two fragments represent two types of valence-electron bonding model (model [I] and model [II]). Therefore, the R-E13 moiety and the R-P component have two and four valence electrons, respectively. The computational results show that the ground states of these two units are a singlet for R-E13 ($[\text{R-E13}]^1$) and a triplet for R-P ($[\text{R-P}]^3$). Therefore, model [I] in **Figure 1** is considered as $[\text{R-E13}]^1 + [\text{R-P}]^1 \rightarrow [\text{R-E13}{\equiv}\text{P-R}]^1$ and model [II] is given as $[\text{R-E13}]^3 + [\text{R-P}]^3 \rightarrow [\text{R-E13}{\equiv}\text{P-R}]^1$.

If the excitation energy (ΔE1) from the triplet ground state to the singlet excited state for R-P is smaller than that for R-E13, then model [I] can be used to interpret the bonding character of RE13≡PR. That is, model [I] demonstrates that the triple bond in RE13≡PR is a single donor-acceptor (E13 → P) σ bond and two donor-acceptor (E13 ← P) π bonds. Therefore, the bonding character of RE13≡PR can be viewed as RE13⇆PR. However, if the promotion energy (ΔE2) from the singlet ground state to the triplet excited state for R-E13 is smaller than that for R-P, then model [II] can be used to explain the bonding character of RE13≡PR. Namely, model [II] shows that the triple bond in RE13 ≡ PR is a single traditional σ bond, a single traditional π bond and a single donor-acceptor (E13 ← P) π bond, so its bonding character can be viewed as RE13⇇PR.

From model [I] and model [II] shown in **Figure 1**, two points need to be emphasized here. First, it is experimentally known that the covalent radius decreases as: Tl (148 pm) > In (142 pm) > Ga (122 pm) > Al (121 pm) > P(107 pm) > B (84 pm) [33]. Therefore, a large difference in the atomic radius results in a significant reduction in the overlap populations between E13 and phosphorus. Consequently, the bonding strength between phosphorus and the E13 element in the heteroatomic analogues of acetylene (RE13≡PR) should be weak. Second, the π bond in the RE13 ≡ PR species is also attributed to the lone pair of the R-P moiety, which is donated into the empty p-π orbital of the R-E13 unit. Since the lone pair of the R-P component

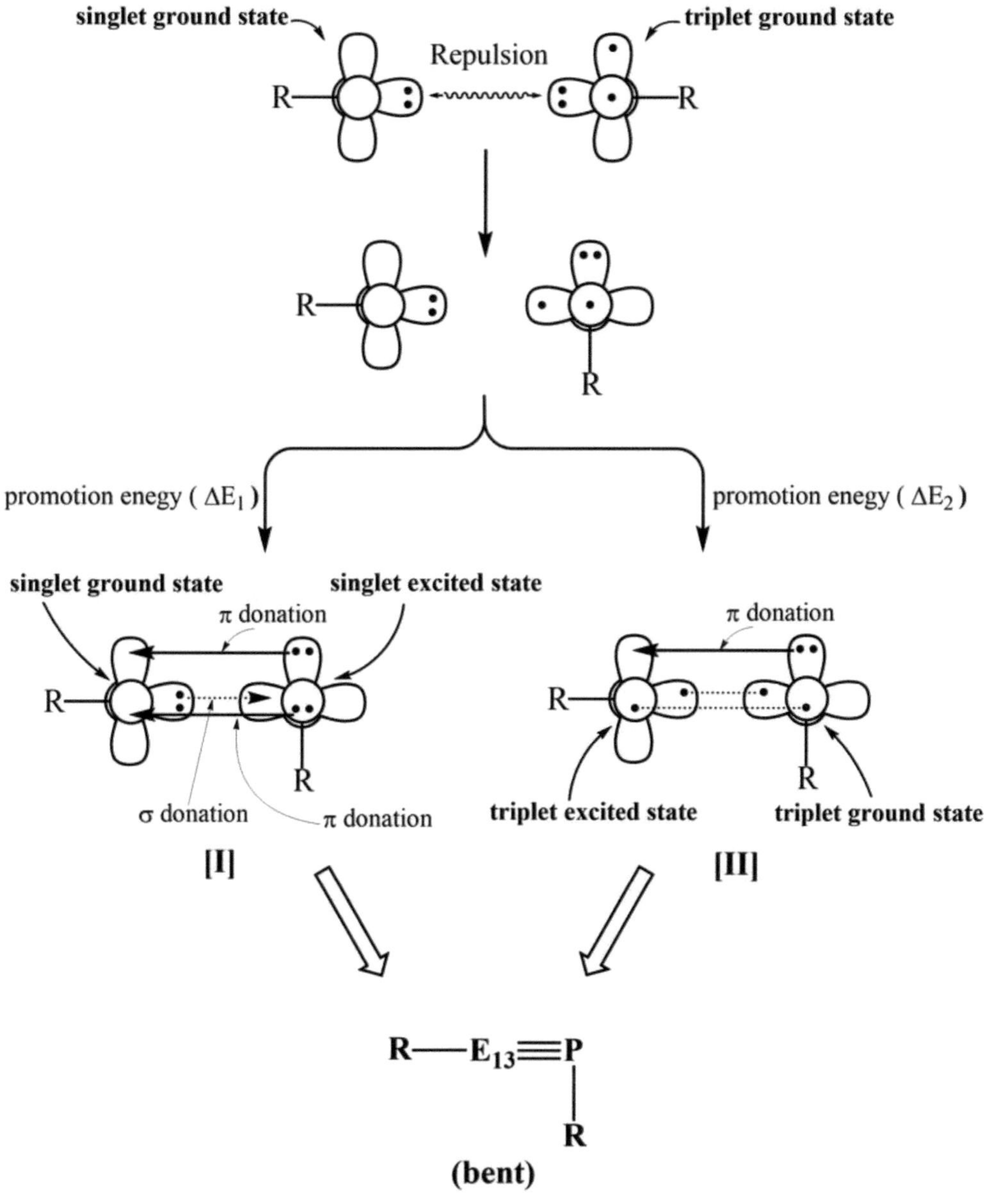

Figure 1. The valence-bond bonding models ([I] and [II]) for the triply bonded RE13 ≡ PR compound.

contains the s valence orbital of phosphorus and the p valence orbital of phosphorus is not the same size as that of the E13 atom, the overlap in the orbital populations between the P and E13 elements is small. In other words, on the basis of the bonding models that are shown in **Figure 1**, the triple bond between E13 and phosphorus is predicted to be very weak.

The computational evidences for these predictions are given in the following sections.

2. Results and discussion

2.1. Small ligands on substituted RE13 ≡ PR

Five small substituents (R), including F, OH, H, CH3 and SiH3, are initially chosen for this study. Three types of density functional theory (DFT) (M06-2X/Def2-TZVP, B3PW91/Def2-TZVP and B3LYP/LANL2DZ + dp) are used to determine the relative stability of the triply bonded RE13≡PR species and its corresponding doubly bonded isomers (R2E13 = P: and: E13 = PR2). In other words, two types of the 1,2-substituent-shift reactions (RE13≡PR → TS1 → R2E13 = P: and RE13≡PR → TS2 →: E13 = PR2) are studied. The respective computational results for RB≡PR [28], RAl≡PR [29], RGa≡PR [30], RIn≡PR [31], and RTl≡PR [32] are schematically shown in **Figures 2–6**.

The computational results that are shown in **Figures 2–6** show that regardless of the type of small substituent that is chosen, the triply bonded RE13 ≡ PR compound cannot be stabilized on the 1,2-migration energy surfaces. That is to say, it is easy for the RE13PR species to migrate to the corresponding doubly bonded R2E13 = P: or: E13 = PR2 isomers rather than to the triply bonded RE13 ≡ PR molecules. The theoretical evidence strongly suggests that the experimental detection of RE13≡PR that features small groups is very unlikely so they are not discussed in this section [28–32].

2.2. Large ligands on substituted R′E13 ≡ PR′

Four bulky groups (R′) are used to study the effects of substituents on the triply bonded RE13≡PR molecules. These are SiMe(Si*t*Bu3)2, Si*i*PrDis2, Tbt (C6H2–2,4,6-{CH(SiMe3)2}3) and Ar* (C6H3–2,6-(C6H2–2,4,6-*i*-Pr3)2) [34, 35]. In order to avoid the London dispersion forces [36], the dispersion-corrected M06-2X/Def2-TZVP level of theory [37] is used to compute geometrical parameters and some properties. The respective results for RB≡PR [28], RAl≡PR [29], RGa≡PR [30], RIn≡PR [31], and RTl≡PR [32] are shown in **Tables 1–5**. The same level of theory is also used to determine the feasibility of producing triply bonded R′E13 ≡ PR′ compounds (**Scheme 1** and **Tables 1–5**).

1. For bulky groups (R′), the E13≡P triple bond distances (Å) are anticipated to be in the range, 1.736–2.023 (B≡P), 2.152□2.183 (Al≡P), 2.146–2.183 (Ga≡P), 2.215–2.362 (In≡P) and 2.336–2.386 (Tl≡P).

2. The computed reaction enthalpies (ΔH1 and ΔH2) that are shown in **Scheme 1** and **Tables 1–5** show that regardless of the bulky ligand that is chosen, the energy of the triply bonded R′E13 ≡ PR′ species is much lower than those of its corresponding doubly bonded R′2E13 = P: or: E13 = PR′2 isomers. This computational evidence indicates that sterically congested ligands kinetically stabilize the triply bonded R′E13 ≡ PR′ compound.

3. The theoretical data in **Tables 1–5** show that the R′-E13 moiety has a singlet ground state, but the R′-P component has a triplet ground state. The production of the triply bonded

http://dx.doi.org/10.5772/intechopen.77055

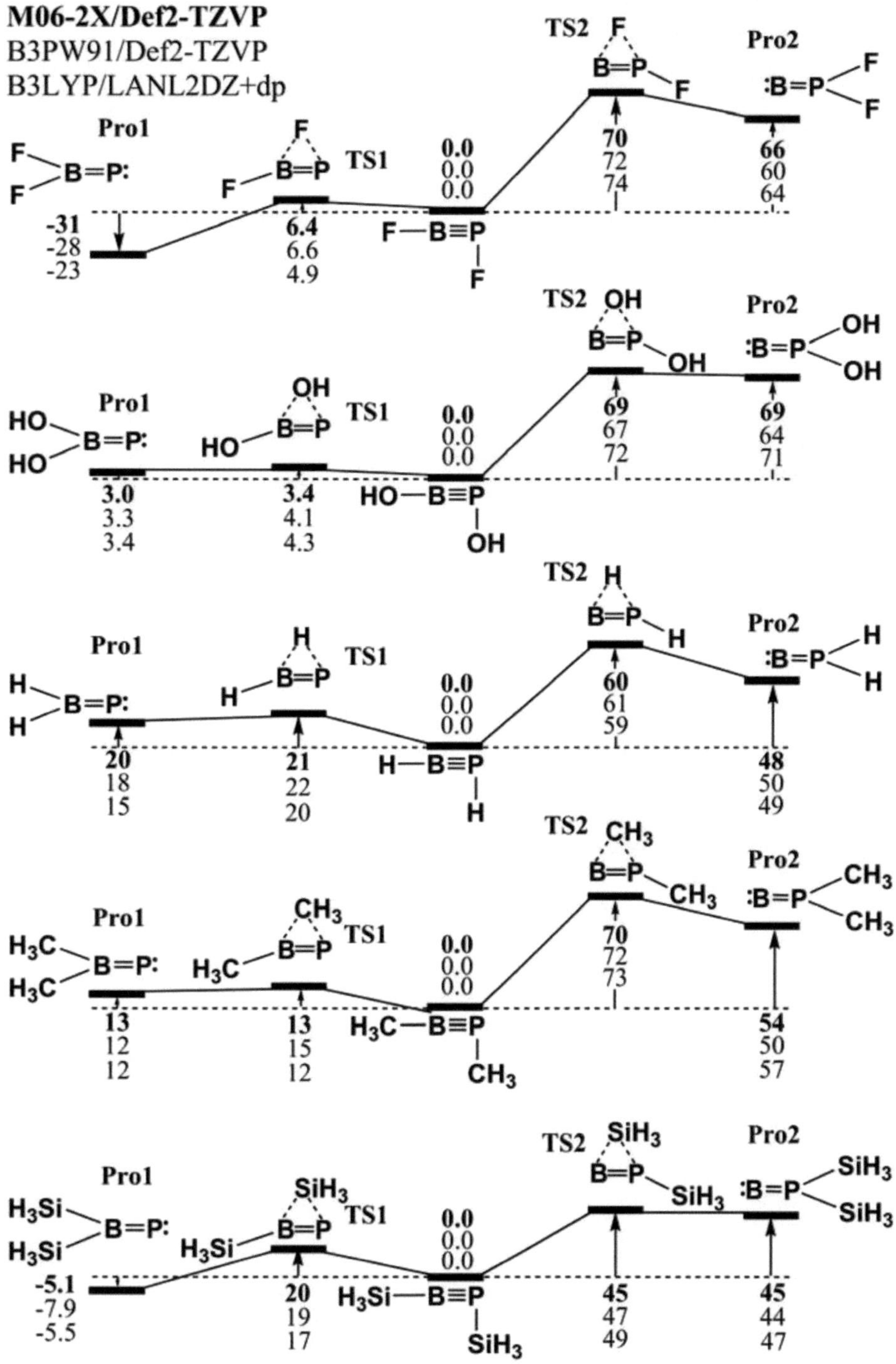

Figure 2. The relative Gibbs free energy surfaces for RB ≡ PR (R = H, F, OH, SiH3, and CH3). These energies are calculated in kcal/mol and are calculated at the M06-2X/Def2-TZVP, B3PW91/Def2-TZVP, and B3LYP/LANL2DZ + dp levels of theory.

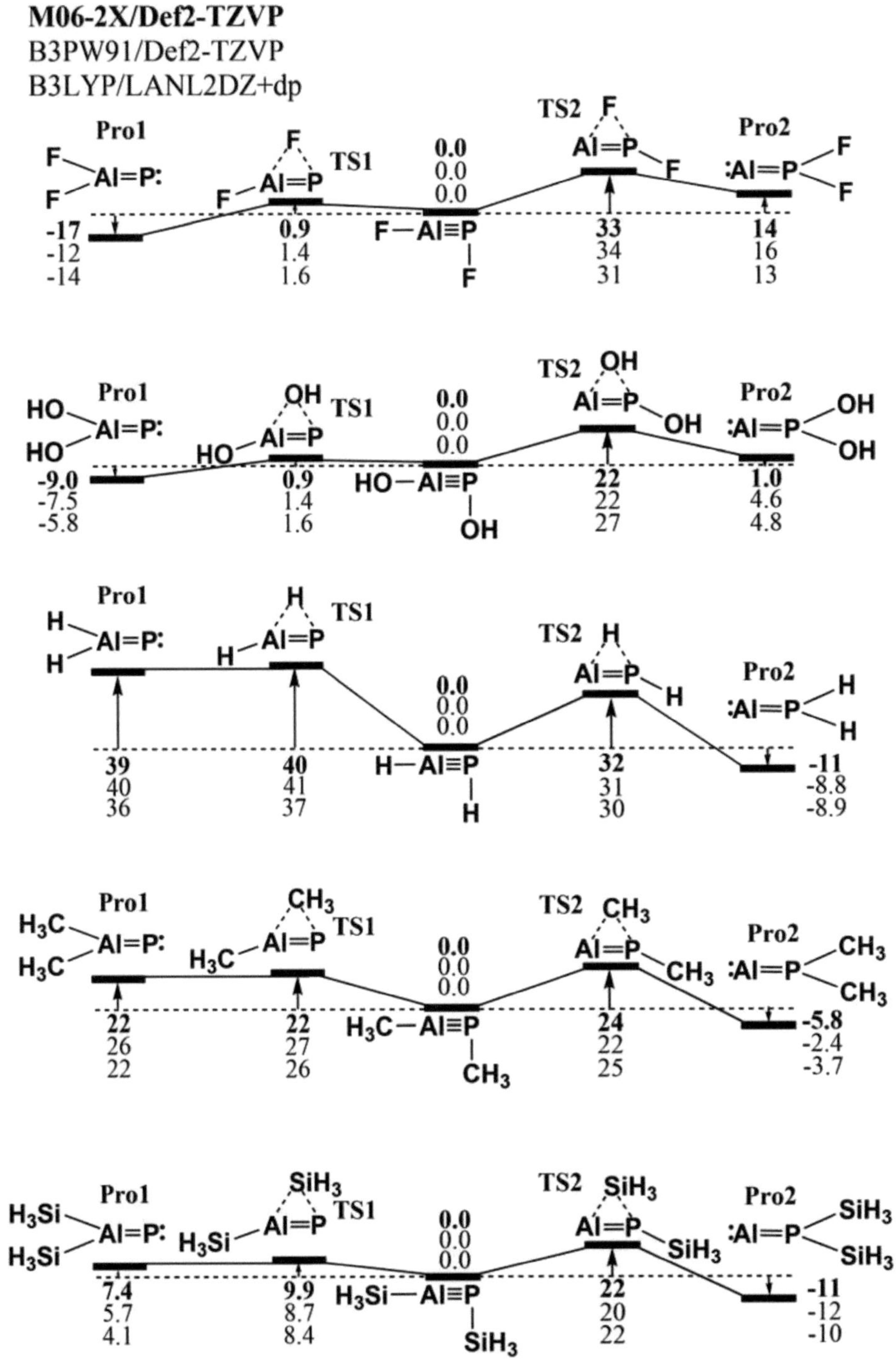

Figure 3. The relative Gibbs free energy surfaces for RAl ≡ PR (R = H, F, OH, SiH3, and CH3). These energies are calculated in kcal/mol and are calculated at the M06-2X/Def2-TZVP, B3PW91/Def2-TZVP, and B3LYP/LANL2DZ + dp levels of theory.

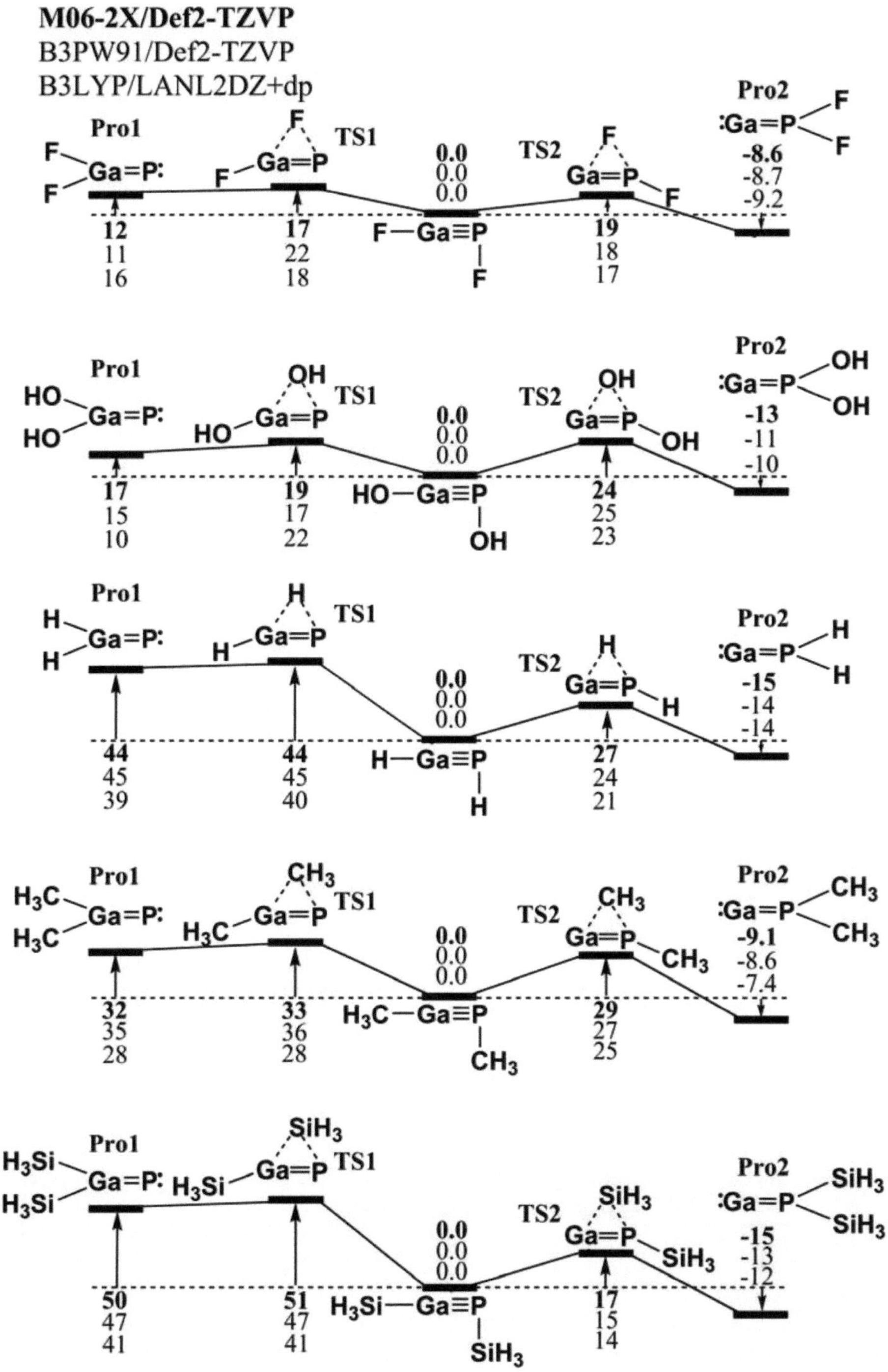

Figure 4. The relative Gibbs free energy surfaces for RGa ≡ PR (R = H, F, OH, SiH3, and CH3). These energies are calculated in kcal/mol and are calculated at the M06-2X/Def2-TZVP, B3PW91/Def2-TZVP, and B3LYP/LANL2DZ + dp levels of theory.

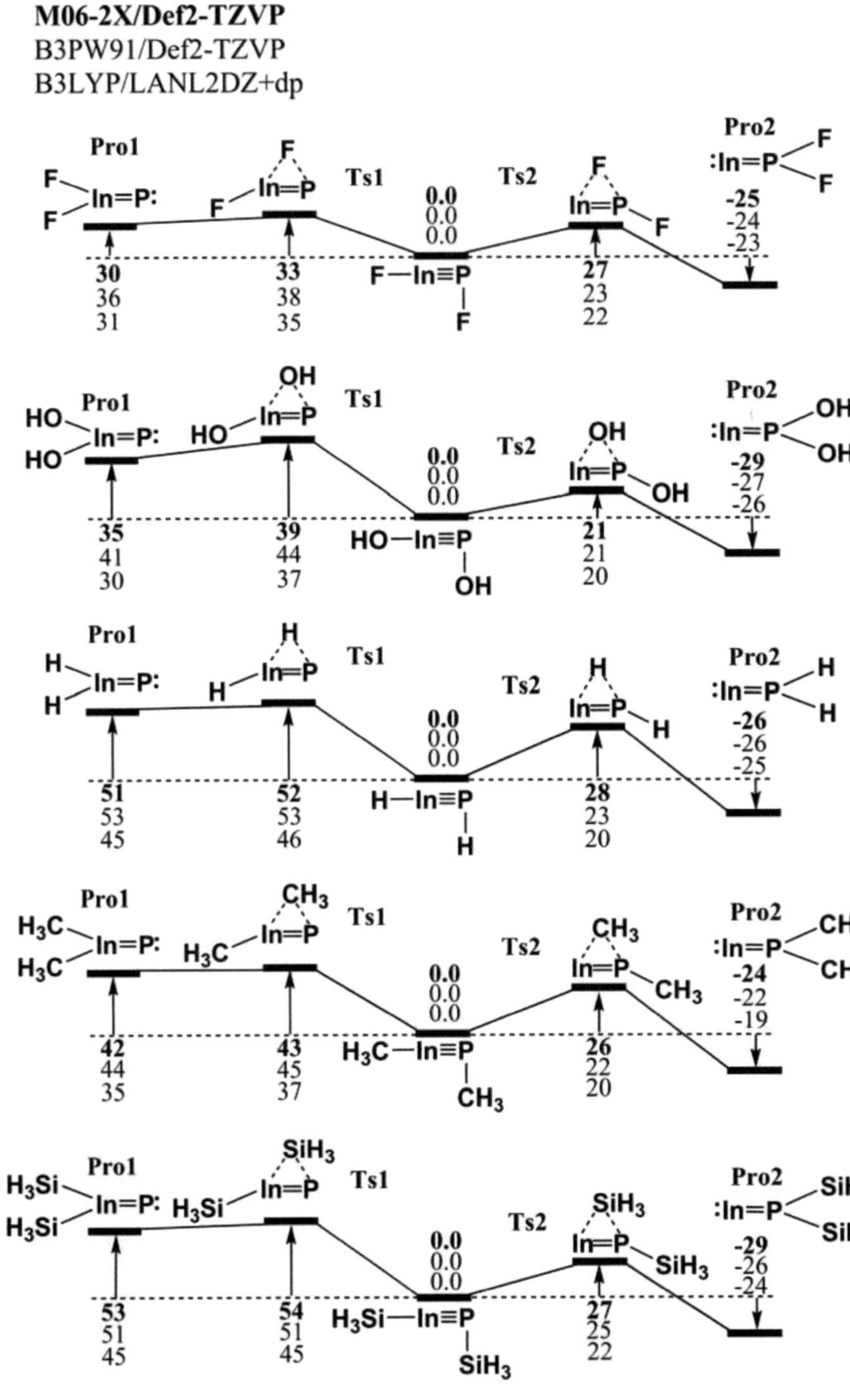

Figure 5. The relative Gibbs free energy surfaces for RIn ≡ PR (R = H, F, OH, SiH3, and CH3). These energies are calculated in kcal/mol and are calculated at the M06-2X/Def2-TZVP, B3PW91/Def2-TZVP, and B3LYP/LANL2DZ + dp levels of theory.

http://dx.doi.org/10.5772/intechopen.77055

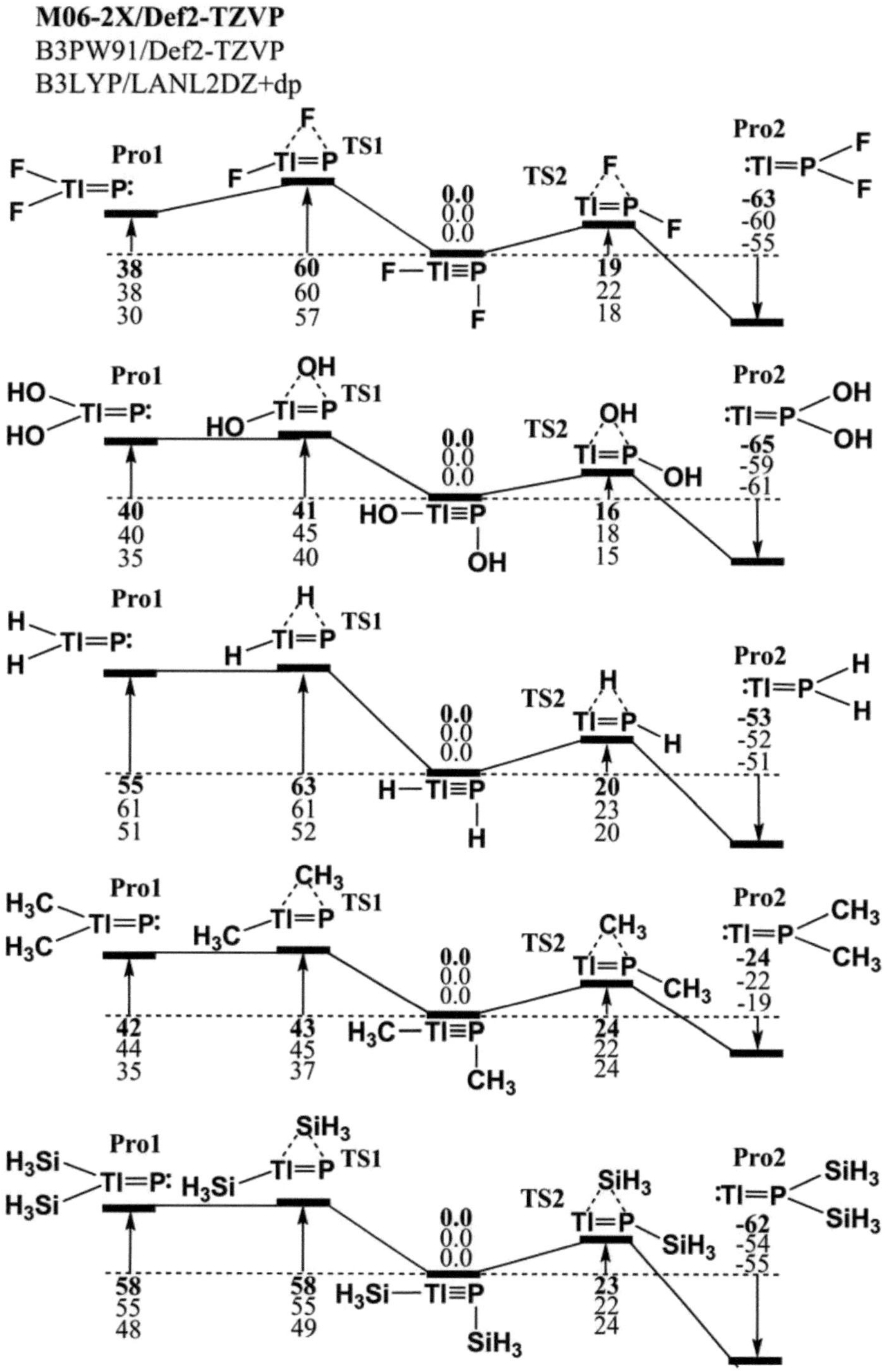

Figure 6. The relative Gibbs free energy surfaces for RTl ≡ PR (R = H, F, OH, SiH3, and CH3). These energies are calculated in kcal/mol and are calculated at the M06-2X/Def2-TZVP, B3PW91/Def2-TZVP, and B3LYP/LANL2DZ + dp levels of theory.

R′E13 ≡ PR′ compound at the singlet ground state constitutes a combination of two triplet units, [R′-E13][3] and [R′-P][3]. Therefore, using the information in **Figure 1**, the bonding nature of the E13 ≡ P triple bond in R′E13 ≡ PR′ can be regarded as RE13⇆PR.

4. The theoretical analyses in Section II shows that the bond order for the E13 ≡ P triple bond should be very weak. **Tables 1–5** show that the Wiberg bond indices (WBI) [28, 39] for RE13≡PR compounds that feature sterically bulky substituents are all a little greater than 2.0. The theoretical evidence demonstrates that RE13≡PR that features bulky groups has only a weak triple bond because the WBI for the C≡C bond in acetylene is computed to be 2.99.

The results of this study show that successful schemes for the synthesis and isolation of triply bonded RE13≡PR molecules are imminent.

R′	**SiMe(Si*t*Bu$_3$)$_2$**	**Si*i*PrDis$_2$**	**Tbt**	**Ar***
B ≡ P(Å)	1.736	2.021	2.023	2.021
∠R′-B-P (°)	157.2	166.0	164.4	166.6
∠B-P-R′ (°)	122.0	112.5	121.3	123.3
∠R′-B-P-R′ (°)	174.7	165.5	168.9	169.5
Q_B′[1]	−0.2574	−0.1395	0.2718	0.3520
Q_P′[2]	−0.1824	−0.3922	0.2260	0.2522
ΔEB′ for R′-B (kcal/mol)[3]	25.92	24.86	28.76	34.64
ΔEP′ for R′-P (kcal/mol)[4]	−33.10	−37.47	−29.74	−30.52
HOMO-LUMO (kcal/mol)	73.76	43.44	47.10	41.60
BE (kcal/mol)[5]	89.54	90.37	85.42	71.43
ΔH_1 (kcal/mol)[6]	73.75	86.65	87.89	87.59
ΔH_2 (kcal/mol)[6]	80.53	77.67	101.7	88.01
WBI[7]	2.388	2.152	1.963	1.966

[1]The natural charge density on the boron atom.
[2]The natural charge density on the phosphorus atom.
[3]ΔEB′ (kcal mol^{-1}) = E(triplet state for R′-B)–E(singlet state for R′-B).
[4]ΔEP′ (kcal mol^{-1}) = E(triplet state for R′-P)–E(singlet state for R′-P).
[5]BE (kcal mol^{-1}) = E(triplet state for R′-B) + E(triplet state for R′-P)–E(singlet for R′B ≡ PR′).
[6]See **Scheme 1**.
[7]The Wiberg bond index (WBI) for the B≡P bond: see references [38, 39].

Table 1. The bond lengths (Å), bond angles (°), singlet-triplet energy splitting (ΔEB′ and ΔEP′), natural charge densities (QB′ and QP′), binding energies (BE), the Wiberg bond index (WBI), HOMO-LUMO energy gaps, and some reaction enthalpies for R′B ≡ PR′ at the M06-2X/Def2-TZVP level of theory.

R′	SiMe(Si*t*Bu$_3$)$_2$	Si*i*PrDis$_2$	Tbt	Ar*
Al ≡ P(Å)	2.168	2.152	2.183	2.175
∠R′-Al-P (°)	166.5	163.4	165.0	167.3
∠Al-P-R′ (°)	117.4	119.7	122.1	121.3
∠R′-Al-P-R′ (°)	166.4	163.8	168.5	167.5
Q_{Al}'[1]	0.9712	0.9210	1.1072	1.326
Q_P'[2]	−0.8751	−0.9674	−0.3430	−0.359
ΔEAl′ for Al-R′ (kcal/mol)[3]	28.89	29.30	42.50	40.22
ΔEP′ for P-R′ (kcal/mol)[4]	−23.10	−27.47	−30.51	−28.52
HOMO-LUMO (kcal/mol)	52.74	34.83	49.98	57.15
BE (kcal/mol)[5]	43.49	54.96	47.51	35.41
ΔH_1 (kcal/mol)[6]	95.15	85.23	91.83	85.60
ΔH_2 (kcal/mol)[6]	96.13	82.75	90.56	85.31
WBI[7]	1.572	1.592	1.685	1.534

[1]The natural charge density on the aluminum atom.
[2]The natural charge density on the phosphorus atom.
[3]ΔEAl′ (kcal mol^{-1}) = E(triplet state for R′-Al)–E(singlet state for R′-Al).
[4]ΔEP′ (kcal mol^{-1}) = E(triplet state for R′-P)–E(singlet state for R′-P).
[5]BE (kcal mol^{-1}) = E(triplet state for R′-Al) + E(triplet state for R′-P)–E(singlet for R′Al≡PR′).
[6]See **Scheme 1**.
[7]The Wiberg bond index (WBI) for the Al≡P bond: see reference [38, 39].

Table 2. The bond lengths (Å), bond angels (°), natural charge densities (QAl′ and QP′), singlet-triplet energy splitting for Al-R′ and P-R′ units (ΔEAl′ and ΔEP′), binding energies (BE), HOMO-LUMO energy gaps, Wiberg bond index (WBI), and some reaction enthalpies for R′Al ≡ PR′ at the dispersion-corrected M06-2X/Def2-TZVP level of theory.

R′	SiMe(Si*t*Bu$_3$)$_2$	Si*i*PrDis$_2$	Tbt	Ar*
Ga ≡ P (Å)	2.167	2.146	2.172	2.183
∠R′-Ga-P (°)	158.2	161.3	152.0	158.4
∠Ga-P-R′ (°)	127.8	120.4	117.3	126.1
∠R′-Ga-P-R′ (°)	176.0	175.5	169.4	166.9
Q_{Ga}'[1]	0.8023	0.8266	0.8952	0.9003
Q_P'[2]	−0.7655	−0.7473	−0.8662	−0.8825
ΔE_{ST} for Ga-R′ (kcal/mol)[3]	30.71	31.34	34.08	38.35
ΔE_{ST} for P-R′ (kcal/mol)[4]	−23.10	−27.47	−23.51	−20.52
HOMO-LUMO (kcal/mol)	83.14	81.83	73.50	71.34
BE (kcal/mol)[5]	91.53	102.9	85.34	89.46
ΔH_1 (kcal/mol)[6]	89.11	94.82	86.31	98.94

R′	SiMe(Si*t*Bu$_3$)$_2$	Si*i*PrDis$_2$	Tbt	Ar*
ΔH_2 (kcal/mol)[6]	86.43	85.91	88.53	84.08
WBI[7]	2.228	2.235	2.017	2.114

[1]The natural charge density on the gallium atom.
[2]The natural charge density on the phosphorus atom.
[3]ΔE_{ST} (kcal mol^{-1}) = E(triplet state for R′-Ga)–E(singlet state for R′-Ga).
[4]ΔE_{ST} (kcal mol^{-1}) = E(triplet state for R′-P)–E(singlet state for R′-P).
[5]BE (kcal mol^{-1}) = E(triplet state for R′-Ga) + E(triplet state for R′-Ga)–E(singlet for R′Ga ≡ PR′).
[6]See **Scheme 1**.
[7]The Wiberg bond index (WBI) for the Ga≡P bond: see reference [38, 39].

Table 3. The bond lengths (Å), bond angels (°), natural charge densities (QGa′ and QP′), singlet-triplet energy splitting (Δ*EST*), binding energies (BE), the HOMO-LUMO energy gaps, the Wiberg bond index (WBI), and some reaction enthalpies for R′Ga ≡ PR′ at the dispersion-corrected M06-2X/Def2-TZVP level of theory.

R′	SiMe(Si*t*Bu$_3$)$_2$	Si*i*PrDis$_2$	Tbt	Ar*
InαP(Å)	2.362	2.337	2.215	2.238
∠R′-In-P (°)	169.6	175.0	177.9	171.4
∠In-P-R′ (°)	115.0	112.0	113.2	115.1
∠R′-In-P-R′ (°)	177.5	172.47	175.4	172.3
Q_{In}'[1]	1.1046	0.9396	0.9489	0.9553
Q_P'[2]	−0.9546	−0.9363	−0.8560	−0.6715
ΔE_{ST} for In-R′ (kcal/mol)[3]	33.93	29.53	22.48	28.41
ΔE_{ST} for P-R′ (kcal/mol)[4]	−28.51	−27.58	−25.64	−22.31
HOMO-LUMO (kcal/mol)	74.96	72.41	87.56	88.43
BE (kcal/mol)[5]	86.51	84.30	92.61	90.64
ΔH_1 (kcal/mol)[6]	92.07	90.08	97.41	87.46
ΔH_2 (kcal/mol)[6]	88.35	89.18	89.26	79.32
WBI[7]	2.263	2.251	2.188	2.174

[1]The natural charge density on the central indium atom.
[2]The natural charge density on the central phosphorus atom.
[3]ΔE_{ST} (kcal mol^{-1}) = E(triplet state for R′-In)–E(singlet state for R′-In).
[4]ΔE_{ST} (kcal mol^{-1}) = E(triplet state for R′-P)–E(singlet state for R′-P).
[5]BE (kcal mol^{-1}) = E(triplet state for R′-In) + E(triplet state for R′-P)–E(singlet for R′In ≡ PR′).
[6]See **Scheme 1**.
[7]The Wiberg bond index (WBI) for the In≡P bond: see reference [38, 39].

Table 4. The bond lengths (Å), bond angels (°), singlet-triplet energy splitting (Δ*EST*), natural charge densities (QIn′ and QP′), binding energies (BE), the HOMO-LUMO energy gaps, the Wiberg bond index (WBI), and some reaction enthalpies for R′In≡PR′ at the B97-D3/LANL2DZ + dp level of theory.

R′	SiMe(Si*t*Bu$_3$)$_2$	Si*i*PrDis$_2$	Tbt	Ar*
Tl ≡ P(Å)	2.386	2.384	2.385	2.336
∠R′–Tl-P (°)	166.9	166.4	168.9	161.2
∠Tl–P–R′ (°)	122.3	113.7	116.2	115.6
∠R′–Tl–P–R′ (°)	171.4	179.5	173.9	174.4
Q_{Tl}'[1]	0.975	0.739	1.166	1.218
Q_P'[2]	−0.860	−0.826	−0.344	−0.257
ΔE_{ST} for Tl–R′ (kcal/mol)[3]	35.91	35.52	31.27	30.24
ΔE_{ST} for P–R′ (kcal/mol)[4]	−43.10	−37.47	−39.74	−40.52
HOMO-LUMO (kcal/mol)	71.27	27.21	58.05	39.34
BE (kcal/mol)[5]	80.24	85.43	62.51	67.89
ΔH_1 (kcal/mol)[6]	91.34	90.49	89.22	87.11
ΔH_2 (kcal/mol)[6]	73.98	72.83	71.27	74.01
WBI[7]	2.116	2.273	2.127	2.201

[1]The natural charge density on the central thallium atom.
[2]The natural charge density on the central phosphorus atom.
[3]ΔE_{ST} (kcal mol^{-1}) = E(triplet state for R′-Tl)–E(singlet state for R′-Tl).
[4]ΔE_{ST} (kcal mol^{-1}) = E(triplet state for R′-P)–E(singlet state for R′-P).
[5]BE (kcal mol^{-1}) = E(triplet state for R′-Tl) + E(singlet state for R′-P)–E(singlet for R′Tl ≡ PR′).
[6]See **Scheme 1**.
[7]The Wiberg bond index (WBI) for the Tl≡P bond: see reference [38, 39].

Table 5. The bond lengths (Å), bond angels (°), singlet-triplet energy splitting (ΔEST), natural charge densities (QTl′ and QP′), binding energies (BE), the HOMO-LUMO energy gaps, the Wiberg bond index (WBI), and some reaction enthalpies for R′Tl ≡ PR′ at the dispersion-corrected M06-2X/Def2-TZVP level of theory.

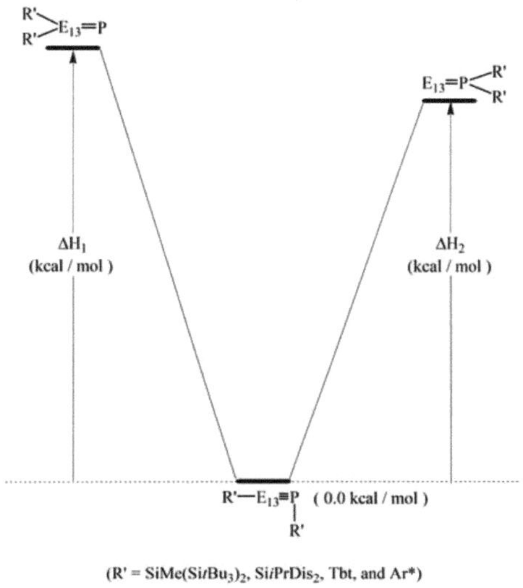

Scheme 1. Several important conclusions can be drawn from the results in **Tables 1–5**.

Acknowledgements

The authors are grateful to the National Center for High-Performance Computing of Taiwan for generous amounts of computing time, and the Ministry of Science and Technology of Taiwan for the financial support.

Author details

Jia-Syun Lu[1], Ming-Chung Yang[1] and Ming-Der Su[1,2]*

*Address all correspondence to: midesu@mail.ncyu.edu.tw

1 Department of Applied Chemistry, National Chiayi University, Chiayi, Taiwan

2 Department of Medicinal and Applied Chemistry, Kaohsiung Medical University, Kaohsiung, Taiwan

References

[1] Nyhlen J, Privalov T. "Frustration" of orbital interactions in Lewis base/Lewis acid adducts: A computational study of H_2 uptake by phosphanylboranes $R_2P{=}BR'_2$. European Journal of Inorganic Chemistry. 2009;**15**:2759-2764

[2] Karsch HH, Hanika G, Huber B, Meindl K, Koöig S, Krüger C, Müller G. Monomeric phosphinoboranes: The role of the B- and P-substituents for P=B multiple bonding. Journal of the Chemical Society, Chemical Communications. 1989;**29**:373-375

[3] Kubo K, Kawanaka T, Tomioka M, Mizuta T. Synthesis and crystal structures of p-iron-substituted phosphinoborane monomers. Organometallics. 2012;**31**:2026-2034

[4] Allen TL, Scheiner AC, Schaefer HF III. Theoretical studies of borylphosphine, its conjugate base, and the lithium salt of its conjugate base. The use of orbital kinetic energies to determine the origin of the driving force for changes in molecular geometry. Inorganic Chemistry. 1990;**29**:1930-1936

[5] Gilbert TM, Bachrach SM. Computational studies of [2+2] and [4+2] pericyclic reactions between phosphinoboranes and alkenes. Steric and electronic effects in identifying a reactive phosphinoborane that should avoid dimerization. Organometallics. 2007;**26**:2672-2678

[6] Privalov T. On the possibility of conversion of alcohols to ketones and aldehydes by phosphinoboranes $R_2PBR'R''$: A computational study. Chemistry-A European Journal. 2009;**15**:1825-1829

[7] Kenward AL, Pierd WE. Heterolytic H_2 activation by nonmetals. Angewandte Chemie International Edition. 2008;**47**:38-41

[8] Coolidge MB, Borden WT. Ab initio calculations on borylphosphines: Prediction of a synergistic substituent effect in diborylphosphine. Journal of the American Chemical Society. 1990;**112**:1704-1706

[9] Geier SJ, Gilbert TM, Stephan DW. Activation of H_2 by phosphinoboranes $R_2PB(C_6F_5)_2$. Journal of the American Chemical Society. 2008;**130**:12632-12633

[10] Paetzold P. The p-centered chemistry of [(di-tert-butylphosphanyl)imino]-(2,2,6,6-tetramethylpiperidino)borane. Advances in Inorganic Chemistry. 1987;**31**:123-170

[11] Grant DJ, Dixon DA. Thermodynamic properties of molecular borane phosphines, alane amines, and phosphine alanes and the $[BH_4^-][PH_4^+]$, $[AlH_4^-][NH_4^+]$, and $[AlH_4^-][PH_4^+]$ salts for chemical hydrogen storage systems from ab initio electronic structure theory. The Journal of Physical Chemistry A. 2005;**109**:10138-10147

[12] Grant DJ, Dixon DA. σ- and π-bond strengths in main group 3−5 compounds. The Journal of Physical Chemistry A. 2006;**110**:12955-12962

[13] Ditty MJT, Power WP. ^{31}P chemical shielding as a function of bond angle: Phosphorus chemical shielding surface of phosphinoborane $R_2PBR'_2$. Canadian Journal of Chemistryan. 1999;**77**:1951-1961

[14] Jemmis ED, Subramanian G. Torsional barriers in $H_2B{:}XCH_3$ (X = NH, O, PH, and S). The Journal of Physical Chemistry. 1994;**98**:8937-8939

[15] Allen TL, Fink WH. Theoretical studies of borylphosphine and its conjugate base. 2. Internal rotation and inversion transition states. Inorganic Chemistry. 1992;**31**:1703-1705

[16] Magnusson E. Conformational energies and structural relaxation in singly bonded silicon, phosphorus and sulfur-compounds. Australian Journal of Chemistry. 1986;**39**:735-745

[17] Geier SJ, Gilbert TM, Stephan DW. Synthesis and reactivity of the phosphinoboranes $R_2PB(C_6F_5)_2$. Inorganic Chemistry. 2011;**50**:336-344

[18] Stephan DW, Erker G. Frustrated Lewis pairs: Metal-free hydrogen activation and more. Angewandte Chemie International Edition. 2010;**49**:46-76

[19] Schulz S. Group 13/15 organometallic compounds-synthesis, structure, reactivity and potential applications. Advances in Organometallic Chemistry. 2003;**49**:225-439

[20] Pons V, Baker T. Soluble boron-nitrogen high polymers from metal-complex-catalyzed amine borane dehydrogenation. Angewandte Chemie International Edition. 2008;**47**:9600-9602

[21] Clark TJ, Lee K, Manners I. Transition-metal-catalyzed dehydrocoupling: A convenient route to bonds between main-group elements. Chemistry-A European Journal. 2006;**12**:8634-8648

[22] Jaska CA, Bartole-Scott A, Manners I. Metal-catalyzed routes to rings, chains and macromolecules based on inorganic elements. Dalton Transactions. 2003;**21**:4015-4021

[23] Maya L. Boron nitride precursors—A perspective. Applied Organometallic Chemistry. 1996;**10**:175-182

[24] Manners I. Polymers and the periodic table: Recent developments in inorganic polymer science. Angewandte Chemie International Edition in English. 1996;**35**:1602-1621

[25] Whittell GR, Manners I. Synthesis of ammonia borane nanoparticles and the diammoniate of diborane by direct combination of diborane and ammonia. Angewandte Chemie International Edition. 2011;**50**:10288-10289

[26] Hugle T, Hartl M, Lentz D. The route to a feasible hydrogen-storage material: MOFs versus ammonia borane. Chemistry-A European Journal. 2011;**17**:10184-10207

[27] Price AN, Nichol GS, Cowley MJ. Phosphaborenes: Accessible reagents for the synthesis of C-C/P-B isosteres. Angewandte Chemie International Edition. 2017;**56**:9953-9957

[28] Lu J-S, Yang M-C, Su M-D. Triple-bonded boron≡phosphorus molecule: Is that possible? ACS Omega. 2018;**3**:76-85

[29] Lu J-S, Yang M-C, Su M-D. Aluminum-phosphorus triple bonds: Do substituents make Al≡P synthetically accessible? Chemical Physics Letters. 2017;**686**:60-67

[30] Lu J-S, Yang M-C, Su M-D. Triply bonded gallium≡phosphorus molecules: Theoretical designs and characterization. The Journal of Physical Chemistry A. 2017;**121**:6630-6637

[31] Lu J-S, Yang M-C, Su M-D. Triply-bonded indium≡phosphorus molecules: Theoretical designs and characterization. American Chemical Society Omega. 2017;**2**:2813-2826

[32] Lu J-S, Yang M-C, Su M-D. Substituent effects on the stability of thallium and phosphorus triple bonds: A density functional study. Molecules. 2017;**22**:1111-1124

[33] Cordero B, Gómez V, Platero-Prats AE, Revés M, Echeverría J, Cremades E, Barragán F, Alvarez S. Covalent radii revisited. Dalton Transactions. 2008;(21):2832-2838

[34] Kobayashi K, Takagi N, Nagase S. Do bulky aryl groups make stable silicon-silicon triple bonds synthetically accessible? Organometallics. 2001;**20**:234-232

[35] Takagi N, Nagase S. Substituent effects on germanium-germanium and tin-tin triple bonds. Organometallics. 2001;**20**:5498-5500

[36] Liptrot DJ, Power PP. London dispersion forces in sterically crowded inorganic and organometallic molecules. Nature Reviews Chemistry. 2017;**1**:4

[37] Zhao Y, Truhlar DG. Density functionals with broad applicability in chemistry. Accounts of Chemical Research. 2008;**41**:157-167

[38] Wiberg KB. Application of the Pople-Santry-Segal CNDO method to the Cyclopropylcarbinyl and Cyclobutyl cation and to Bicyclobutane. Tetrahedron. 1968;**24**:1083-1096

[39] Reed AE, Curtiss LA, Weinhold F. Intermolecular interactions from a natural bond orbital. Donor-Acceptor Viewpoint. Chemical Reviews. 1998;**88**:899-926

Section 3

Phosphorus Flow and Mitigation Strategies

Chapter 4

Phosphorus Eutrophication and Mitigation Strategies

Lucy Ngatia and Robert Taylor

Additional information is available at the end of the chapter

http://dx.doi.org/10.5772/intechopen.79173

Abstract

Phosphorus (P) eutrophication in the aquatic system is a global problem. With a negative impact on health industry, food security, tourism industry, ecosystem health and economy. The sources of P include both point and nonpoint sources. Controlling P inflow from point sources to aquatic systems have been more manageable, however controlling nonpoint P sources especially agricultural sources remains a challenge. The forms of P include both organic and inorganic. Runoff and soil erosion are the major agents of translocating P to the aquatic system in form of particulate and dissolved P. Excessive P cause growth of algae bloom, anoxic conditions, altering plant species composition and biomass, leading to fish kill, food webs disruption, toxins production and recreational areas degradation. Phosphorus eutrophication mitigation strategies include controlling nutrient loads and ecosystem restoration. Point P sources could be controlled through restructuring industrial layout. Controlling nonpoint nutrient loads need catchment management to focus on farm scale, field scale and catchment scale management as well as employ human intervention which includes ferric dosing, on farm biochar application and flushing and dredging of floor deposits. Ecosystem restoration for eutrophication mitigation involves phytoremediation, wetland restoration, riparian area restoration and river/lake maintenance/restoration. Combination of interventions could be required for successful eutrophication mitigation.

Keywords: agriculture, aquatic, eutrophication, mitigation, phosphorus

1. Introduction

Globally many aquatic ecosystems have been negatively affected by phosphorus (P) eutrophication [1]. Phosphorus is a primary limiting nutrient in both freshwater and marine systems [2, 3]. Phosphorus eutrophication is defined as the over enrichment of aquatic ecosystems with P leading to accelerated growth of algae blooms or water plants, anoxic events, altering biomass and species composition [4–8], eutrophication is a persistent condition of surface waters and a widespread environmental problem. Aquatic systems affected by eutrophication often exhibit harmful algal blooms, which foul water intakes and waterways, fuel hypoxia, disrupt food webs and produce secondary metabolites that are toxic to water consumers and users including zooplantkton, shellfish, fish, domestic pets, cattle and humans [9]. Excess phosphorus inputs to lakes usually come from industrial discharges, sewage and runoff from agriculture, construction sites, and urban areas [1, 6]. Many countries have regulated point sources of nutrients for example industrial and municipal discharges, however, nonpoint sources of nutrients such as runoff from urban and agricultural lands have replaced point sources as the driver of eutrophication in many countries [10].

The plant availability of phosphate fertilizer is reduced by sorption and organic complexation in the soil, therefore fertilizer applications greater than the amount required by the crop are used to counteract the strong binding of the phosphate to the soil matrix leading to increased P content of managed soils [11]. Mining phosphorus (P) and applying it on farm for animal feed and crop production is altering the global P cycle, leading to P accumulation in some of the world's soils [4]. Over application of P fertilizer to soil is in itself wasteful, but the transport of P to aquatic ecosystems by erosion is also causing widespread problems of eutrophication [12, 13]. Using global budget [4] estimated the increase in net P storage in terrestrial and freshwater ecosystems to be at least 75% greater than preindustrial levels of storage. Large portion of P accumulation occurs as a result of increased agriculture intensification, land use change and increased runoff [14–16].

2. Sources of phosphorus

Excess phosphorus inputs to lakes usually come from industrial discharges, sewage and runoff from agriculture, construction sites, and urban areas [6]. The input of P to soil creates the potential for an increase in transfer to the wider environment. Phosphorus sources can be natural which includes indigenous soil P and atmospheric deposition and/or anthropogenic which includes fertilizers, animal feed input to the farm and manure applied to the soil [17]. In addition ground water can potentially contribute significant amount of P in water bodies driving eutrophication process [18].

As a result of P point source control in many countries the nonpoint P source especially agriculture is the main pollutant of aquatic systems [10]. The major source of nonpoint nutrient input to water bodies is the excessive application of fertilizer or manure on farm which cause P accumulation in soils [4]. Phosphorus retention has been caused by an excess of fertilizer and animal feed inputs over outputs of agricultural products [19]. For example [12] indicated that less than 20% of P input to the Lake Okeechobee watershed in fertilizers was output in agricultural as well as other products.

3. Forms of phosphorus

Most studies on plant nutrition often consider only inorganic P to be biologically available, however organic phosphorus is abundant in soils and its turnover can account for the majority of P taken up by vegetation [20, 21]. Soil phosphorus exists in a range of organic and inorganic compounds that differ significantly in their biological availability in the soil environment [22]. The inorganic P compounds mainly couple with amorphous and crystalline forms of Al, Fe, and Ca [23] which is highly influenced by soil pH [24]. The organic phosphorus in most soils is dominated by a mixture of phosphate monoesters (e.g., mononucleotides, inositol phosphates) and phosphate diesters (mainly nucleic acids and phospholipids), with smaller amounts of phosphonates (compounds with a direct carbon-phosphorus bond) and organic polyphosphates (e.g., adenosine triphosphate) [25]. Plants can manipulate their acquisition of P from organic compounds through various mechanisms, some of the mechanisms allow plants to utilize organic P as efficiently as inorganic phosphate [26, 27]. In lake sediment in China, the heavily polluted sediments contained higher organic P fractions compared to moderately and no polluted sediments [28]. Increased pH can alter the availability of P binding sites on ferric complexes as a result of competition between hydroxyl ions and bound phosphate ions [29]. Anoxic condition leads to release of P as a result of reduction of ferric to ferrous iron [30]. In addition presence of sulfate could lead to reaction of ferric iron with sulfate and sulfide to form ferrous iron and iron sulfide leading to release of P [31]. Increased temperature can reduce adsorption of P by mineral complexes in the sediment [32]. Other physicochemical processes affecting release of P from the sediment include temperature, pH potential, redox, reservoir hydrology and environmental conditions [33]. These physical chemical processes are further complicated by the influence of biological processes for example mineralization, leading to a complex system governing the release of P across sediment water interface [33].

4. Translocation/mobilization of phosphorus

Phosphorus in most cases enters aquatic ecosystems sorbed to soil particles that are eroded into rivers, lakes and streams [34, 35]. Watershed land use and P concentration in watershed soils strongly influence potential P pollution of aquatic ecosystems [4]. In addition any factor accelerating erosion or elevating P concentration in the soil increases the potential P runoff to aquatic systems [34, 35]. Mobilization of P involves biological, chemical and biochemical processes. The processes are grouped into solubilization or detachment mechanisms and are defined by the physical size of the P compounds that are mobilized [17] and it has been indicated that the potential for solubilization increases with increasing concentrations for extractable P. However, organic P has an important but little understood role in determining P solubilization [25]. Detachment of soil particles and associated P is mainly linked to soil erosion, which provides a physical mechanism for mobilizing P from soil into waters bodies [36].

Depending on site conditions diffuse P transport occurs as particulate or dissolved P in overland flow, channelized surface runoff, drainage, or groundwater [18]. In ground water

P concentration is considered to be low [37]. This is as a result of orthophosphate P being adsorbed in the soil and sediment in the vadose or the saturated zone [18]. However, wastewater has been reported to cause heavy groundwater contamination leading to P elevation [38, 39].

5. Consequences of eutrophication

Consequences of eutrophication include excessive plant production, blooms of harmful algae, increased frequency of anoxic events, and death of fish, leading to economic losses and health implications which include costs of water purification for human and industry use, losses of fish and wildlife production and losses of recreational amenities [10, 40]. Some of the consequences of eutrophication includes:

5.1. Food/fishing industry

High level of Lake Eutrophication has led to suffocating of fish population on a massive scale with a very negative repercussions on the economy [41]. The total economic loss incurred from 1998 algal bloom in the Lake Tai China catchment area was estimated at U.S.$6.5 billion. During winter of 2002–2003, a severe oxygen deficit induced a fish kill in the eutrophicated two-basin Lake Aimajarvi in southern Finland, which resulted in cascading effects on the lower trophic levels of the lake [42].

5.2. Tourism

Coastal areas are an important economic source for tourism [43]. The algal bloom have degraded the investment environment and damaged the hospitality and tourism industry [41].

5.3. Human/animal healthy

Toxin producing algae can cause mass mortalities of fish marine mammals, birds and human illness through consumption of sea food [44]. It is estimated that 60–80 species of about 400 known phytoplankton are toxin producing and capable of producing harmful algal blooms [8]. In humans, toxins arising from harmful algal blooms have mainly been from shellfish consumption [44], bivalve shellfish have been reported to graze on algae and concentrate toxins effectively. As a result the poisoning can lead to paralytic shellfish poisoning, diarrhetic shellfish poisoning, neurotoxic shellfish poisoning, amnesic shellfish poisoning and azaspiracid shellfish poisoning. In addition there are many respiratory complaints from inhaling contaminated aerosols [45]. A case reported that in July 2002 teenage boys swam in a blue-green algae covered golf course in Dane County, Wisconsin. They all became ill, the most severe symptoms occurred in the boys who swallowed water. Approximately after 48 h one of the boys suffered a seizure and died of heart failure, the coroner identified anatoxin-a as the most likely underlying cause of death [46, 47].

In May and June 1998, over 200 California sea lions died and others displayed signs of neurological dysfunction along the central California coast, this was linked to a harmful algal blooms [48].

5.4. Water quality

Harmful algal blooms is a cause of restriction on drinking water, fisheries and recreation water uses leading to significant economic consequences [49]. The presence of algal bloom and other species have disrupted the normal supply of drinking water in many parts of the world for example China [41]. The presence of algal blooms in Lake Tai severely affected industrial and agricultural production as well as lives of the urban dwellers. Whereby in 1990 algal bloom forced the shutdown of the entire water supply system and triggered a crisis in water security for the urban population. The direct economic loss was estimated at about U.S.$30 million. Harmful algal blooms present significant challenges for achieving water quality protection and restoration goals especially when these toxins confound interpretation of monitoring results and environmental quality standards implementation efforts for other chemicals and stressors [49].

6. Phosphorus eutrophication mitigation strategies

There is need to reduce anthropogenic nutrient inputs to aquatic ecosystems in order to reduce the negative effects of eutrophication [50]. It has been indicated that reducing P input in the water bodies leads to eutrophication mitigation [16, 51] (**Table 1**) [16]. Derived this conclusion from four methods, all long-term studies at ecosystem scales:

i. long-term case histories,

ii. multiyear whole lake experiments,

iii. experiments where chemical treatments are used to remove phosphorus from the water column, and

iv. chemical additions to inhibit return of phosphorus from the sediments to the water column.

However, [6, 52] argued that P based nutrient mitigation commonly regarded as the key tool in eutrophication, in many cases has not yet yielded the desired reductions in water quality and nuisance algal growth in water bodies has not reduced in decades of reducing P input. [52, 53] argued that these observations could be as a result of:

i. legacies of past land-use management;

ii. decoupling of algal growth responses to river P loading in eutrophically impaired aquatic system; and

iii. recovery trajectories, which may be nonlinear and characterized by thresholds and alternative stable states.

Therefore, as a result of these contrasting findings there is need in some cases to consider a combination of different P mitigation strategies for example employing control of nutrient loading, physicochemical and physicomechanical method simultaneously. Control or mitigation of P eutrophication should encompass multiple components which could include; control of pollutant sources, restoration of the damaged ecosystem, and catchment management [41]. The mitigation strategies includes:

Name of water body	Country	Location
Little Otter Lake	Canada	45°N, 80°W
Gravenburst Bay	Canada	45°N, 80°W
Kootenay Lake	Canada	50°N, 117°W
ELA Lakes 226 NE, 303, 304, 261	Canada	50°N, 94°W
Lake Eric	Canada, USA	42°N, 81°W
Lake Ontario	Canada, USA	44°N, 78°W
Lake Huron	Canada, USA	44°N, 82°W
Lake Fure	Denmark	56°N, 12°E
Rhine River	Europe	49°N, 7°E
Schlachtensee	Germany	52°N, 13°E
Lake Tegel	Germany	53°N, 13°E
Lake Balaton	Hungary	47°N, 18°E
Lago Maggiore	Italy	46°N, 9°E
Lake Mjosa	Norway	61°N, 11°E
Lake Vattern	Sweden	48°N, 15°E
Lake Vanern	Sweden	59°N, 13°E
Lake Hjalmaren	Sweden	59°N, 16°E
Lake Malaren	Sweden	59°N, 17°E
Lake Norrviken	Sweden	59°N, 18°E
Murtensee	Switzerland	47°N, 7°E
Lake Lucerne	Switzerland	47°N, 8°E
Turlersee	Switzerland	47°N, 8°E
Hallwilersee	Switzerland	47°N, 8°E
Sempachersee	Switzerland	47°N, 8°E
Lake Zurich	Switzerland	47°N, 9°E
Pfaffikersee	Switzerland	47°N, 9°E
Lake Constance	Switzerland, Austria, Germany	48°N, 9°E
Lake Geneva	Switzerland, France	46°N, 6°E
Sacramento River	USA	40°N, 121°W
Lake onondaga	USA	43°N, 76°W
Lake Michigan	USA	44°N, 87°W
Moses Lake	USA	47°N, 119°W
Lake Washington	USA	48°N, 112°W

Table 1. Examples of fresh waters in some countries where eutrophication decreased following the control of phosphorus inputs. Latitudes and longitudes are given. Lakes recovered by using chemicals to precipitate phosphorus are not included, modified from [16].

6.1. Control of nutrient loads

6.1.1. Restructuring of industrial layout

Point source P originating from mines, factories and residence form one of the most important sources of P to water bodies [41]. For example Lake Tai in China, its catchment area used to be full of heavy industrial polluters, for example, chemical and dye factories. The township lacked adequate facilities for treating waste water before disposal. Therefore, to mitigate pollution coming from the industry it is important to shut down heavily polluting industries. While as the less polluting plants could be relocated to a designated industrial part to ensure centralization and effectiveness in handling pollution control.

6.1.2. *Farm/field/catchment management*

It has been much easier to control point source P, therefore making nutrient discharge from agricultural fields the chief source of pollution [41]. As a result nutrient discharge from agricultural fields could be addressed through farm, field and catchment management or rationalization of land use [41].

6.1.2.1. *Farm scale management*

In the farm scale environmentally sound fertilizer application and nutrient handing is important this would be achieved through appropriate placement and proper timing of application. This would result in moderate P levels in the soils. In Addition P input could be reduced through increasing digestible P in fodder and reducing total P [51] (**Table 2**).

6.1.2.2. *Field scale management*

To avoid transport of particulate P and leaching of P, increase soil storage there is need to change soil management. In addition there is need to change crop management in order to reduce run off and reduce leaching [51] (**Table 3**).

Strategy	Aim	Measure
Change P input	Avoid high P content in fodder by increasing digestible P and lowering total P content in feed	Use feed with a lower content of phytate-P or add phytase to feed to increase digestibility of phytate-P
Change P output	Exploit the commercial value of the manure surplus	Make products for export or for arable farms. Produce secondary P resources for industries by incineration to P-ash
Environmentally sound fertilizer application & nutrient handling	Reduce P content of the soil at high risk hot spots	Don't apply manure and P fertilizer at high risk hot spots
	Increase P efficiency of crop uptake via appropriate placement and time of application	Apply P near the roots instead of broadcast Avoid applying manure and P fertilizers before heavy rainfall or prolonged rainfall Phase nutrient fertilization application over the year
	Incorporate soil P into management strategy to achieve moderate soil P levels	Make use of available P in soils to avoid high risk hot-spots

Table 2. Mitigation strategies for nutrient management at farm scale. Modified from [51].

Strategy	Aim	Measure
Change crop management	Change cropping system	Introduce crop rotation and include more years of grass or develop mixed (perennial and annual) cropping systems
	Avoid erosion and reduce surface runoff	Grassland instead of arable crops or grow deep-rooting crops
	Avoid leaching	Apply catch crops (and harvest the products) Crop production without fertilization (P mining)
Change soil management	Avoid transport of particles or particulate P	No tillage/direct drilling: leaving more than 30% of the soil covered with plant residues or undisturbed stubble Shallow cultivation: Soil tillage to <10 cm depth. No inversion Contour ploughing Switch from autumn tillage to spring tillage Reduce soil compaction and improve soil structure
	Avoid leaching of dissolved P concentrations in soils	Conventional ploughing or interspersing periods of ploughless tillage with conventional ploughing
	Reduce nutrient budgets and increase soil storage capacity by extensification and agro-forestry	Introduce crop rotation and include more years of grass or develop mixed cropping systems with perennial and annual crops

Table 3. Mitigation strategies at field scale modified from [51].

6.1.2.3. Catchment scale management

In the catchment scale eutrophication mitigation strategy would involve water management, land use management and landscape management [51]. Water management could be achieved through reducing runoff flow and avoiding subsurface leaching. Land use management would involve protecting vulnerable areas and improving sink and sources of P by changing agricultural use patterns. Land scape management would include reducing direct sources of P from farmyard, livestock and reducing surface runoff and erosion from field to field within the catchment [51] (**Table 4**).

6.1.3. Human intervention

It has been demonstrated that often the nutrient load and algal blooms in water bodies respond slowly to interventions aimed at controlling external nutrient sources because of the nutrients replenished from waterbodies deposits [54]. As a result P could be reduced through physiochemical and physicomechanical methods. Whereby the P is trapped and removed from the system or trapped on farm and its mobility to aquatic system reduced.

6.1.3.1. Ferric dosing

Reduction in the external P loading for control of algal biomass in the water reservoirs can be achieved by the use of ferric dosing. Ferric dosing technique is a physiochemical method

Strategy	Aim	Measure
Land use management	Changing agricultural use patterns to improve location of sinks and sources of P	Alternate arable land and grassland Avoid certain crops in hilly areas Locate crops with high nutrient uptake on bottom lands
Landscape management	Reduce surface runoff and erosion from field to field within the catchment	Re-site gateways and paths: trails, roads, controlled access for livestock and machinery
	Reduce direct losses from livestock	Prevent contact with surface water: fences, bridges
	Reduce direct losses from farm yards	Minimize volume of dirty water produced and collect farm yard runoff
Water management	Change runoff flow by blocking or reducing overland flow	Create ponding systems Construct grassed waterways Improve surface irrigation
	Avoid subsurface losses through leaching	Remove trenches and ditches or allow to deteriorate Install drains Let drainage water irrigate meadows

Table 4. Mitigation strategies at catchment scale modified from [51].

and involves the addition of ferric sulfate or alternatives to the pumped input, to precipitate dissolved particulate and orthophosphate in the coming water. The system is coupled with filtration to remove the ferric/phosphorus floc [33]. Resulting in a significant reduction of P in the pumped inputs to the aquatic system.

6.1.3.2. Flushing and dredging of floor deposits

Flushing and dredging of floor deposits is a physicomechanical methods meant to remove the already accumulated P from the aquatic system floor [55]. The limitation with both ferric dosing and flushing and dredging of floor deposits is that they provide temporal solutions and do not address the root cause of the problem. Once the intervention is stopped nutrients levels goes back to the former status. However, the success of these methods are dependent on their being implemented together with control of nutrient load intervention.

6.1.3.3. Biochar potential in phosphorus adsorption on farm

Application of P on farm has potential to mitigate P eutrophication though P adsorption leading to reduction in P translocation (**Figure 1**). The P adsorption to biochar is favored by increased biochar pyrolysis temperature and is biochar biomass species specific [24]. The increase of biochar aromatic C (**Figure 2**) and pH adjustment with high biochar pyrolysis temperature is important for P retention [24]. Biochar is a byproduct of biofuel production, therefore increased production of biofuel will be consistent with biochar availability in future.

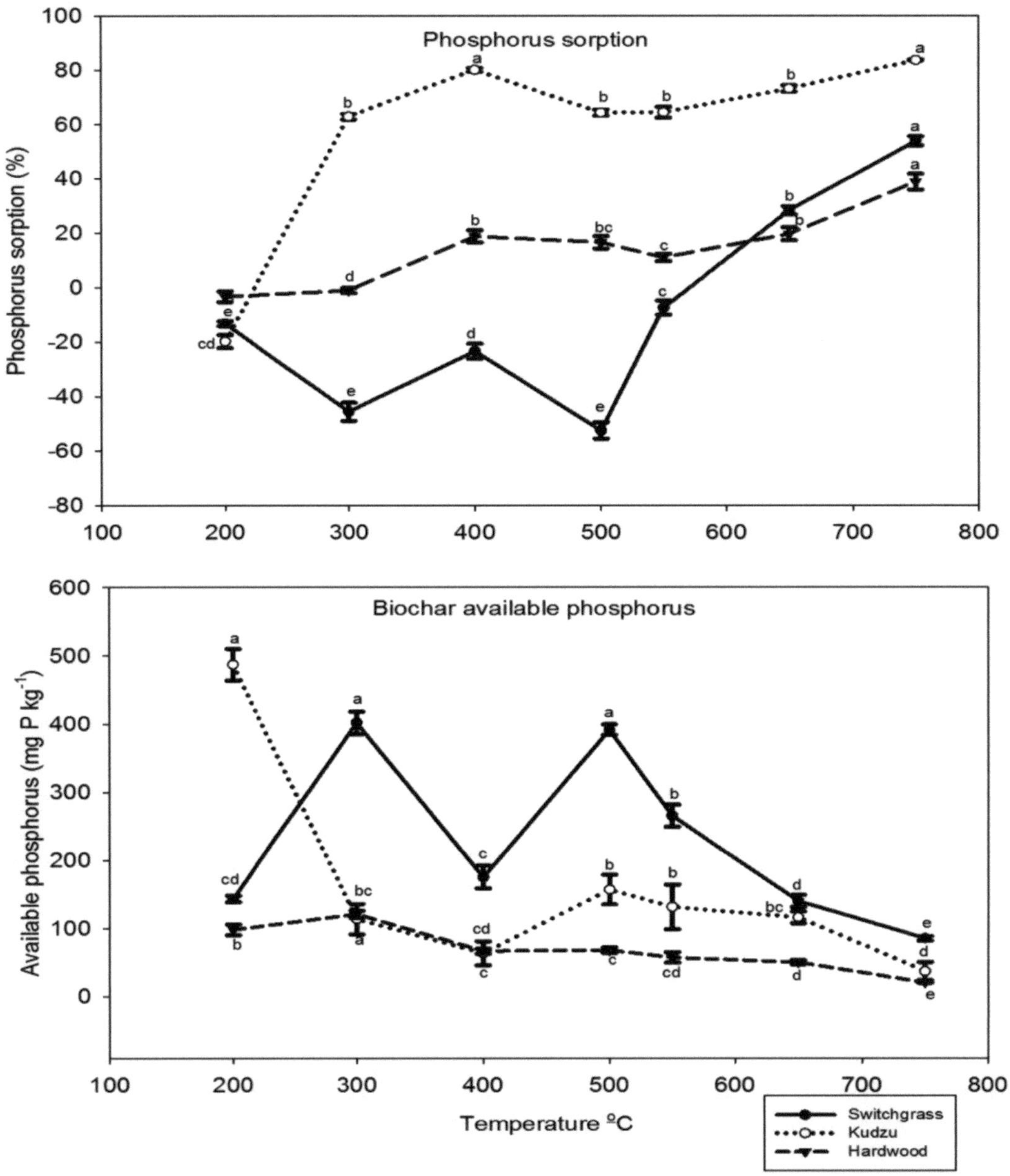

Figure 1. Biochar phosphorus adsorption; from [24].

6.2. Aquatic ecosystem restoration

Ecosystem restoration focus on rehabilitating the functionality of the damaged ecosystem and it relevant physical, chemical and biological properties [41].

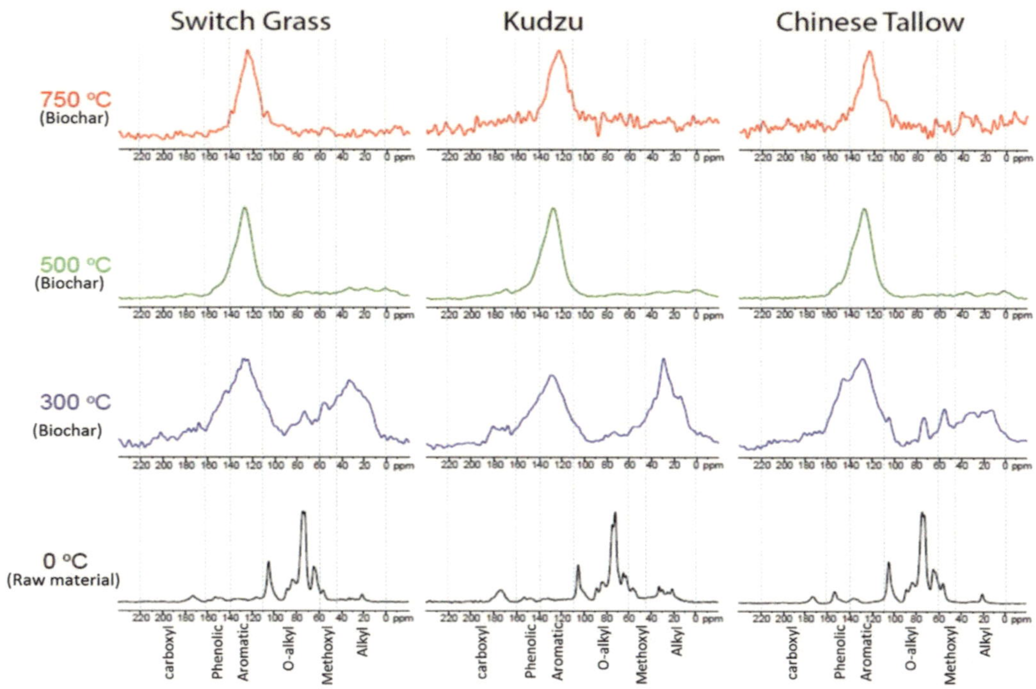

Figure 2. Biochar carbon functional groups as determined by nuclear magnetic resonance (NMR); from [24].

6.2.1. *Phytoremediation*

Ecological restoration may be accomplished through reduction of algae in the water bodies through aquatic plants. Being primary producers, advanced aquatic plants and micro-organisms compete with each other for ecological resources, such as light, nutrients, and living space. During their growth, advanced aquatic plants release chemical substances that are conducive to inhibiting algal production, as well as directly absorbing nitrogen and phosphorus in water. Storage of these elements in the plants means that they can be effectively extracted from the water through physical removal of these plants from the lake, thus reducing the nutrient level in lake waters [56].

6.2.2. *Wetland restoration and constructed wetlands*

Wetland restoration and constructed wetlands retain nutrient loss from upstream fields protecting the aquatic system (**Table 5**) [51]. Wetlands play a key role in P removal due processes that include peat accretion (sorption and burial in soil and sediments), uptake by microbes and vegetation and precipitation by iron and aluminum [57–59].

6.2.3. *Riparian area restoration*

Although [18] indicated that ground water contained elevated P which was a driver of eutrophication, there was no clear evidence of the location and sources of the pollution, as a result

Lake rehabilitation and restoration	Reduce the P concentration of lake water	Control P inlet and prolong residence time of water
River maintenance and restoration	Increase capacity to retain nutrient	Re-meander, restore flood plains and reconnect inundation areas Limit cutting of vegetation and reduce regular removal of gravel and impediments to flow
Wetland restoration and constructed wetlands	Retain nutrient loss from upstream fields in wetlands	Create wetlands in agricultural areas with substantial P losses

Table 5. Mitigation strategies in aquatic ecosystems. From [51].

measure to decrease groundwater derived P loads cannot target the contamination at its source in the catchment. Hence the need to implement measures in the riparian area to eliminate groundwater P directly before it enters the water body.

6.2.4. River/lake maintenance and restoration

River maintenance and restoration is important in increasing nutrient retention capacity. Lake rehabilitation and restoration reduce the P concentration of lake [51] (**Table 5**).

7. Conclusion

Phosphorus eutrophication is a major environmental problem globally resulting in negative impact on the economy, health and tourism sector. Phosphorus eutrophication is caused by both point and nonpoint sources of P. In many countries point source of P has been better controlled while nonpoint sources of P which is mainly agricultural sources have been an ongoing major challenge. The excessive P in the farm is a result of excessive fertilizer and manure application on farm. Phosphorus has mainly been translocated from the source to aquatic systems through run off and leaching and in some isolated cases through ground water. There is need to implement mitigation strategies. This chapter recommends implementation of measures to control nutrient loads through restructuring industrial layout which is a point sources of P pollution. To address nonpoint source of P there is need to implement catchment management measures and ecosystem restoration measures. The catchment management encompasses farm scale, field scale and catchment scale measures that will either reduce P availability for translocation or retain P in the catchment area. Human intervention is equally important to ensure removal of P from already contaminated aquatic system or prevention of P translocation to the water bodies; human intervention includes ferric dosing, flushing and dredging of floor deposits and biochar application on farm. The human intervention especially ferric dosing and flushing and dredging of floor deposits has limitation because once intervention is abandoned the P status could easily revert to pre-intervention status. Previous studies indicate contrasting finding on the success of the mitigation strategies, whereby some reported success while others indicated no response to P eutrophication mitigation. As a result, it seems there is need in some cases to combine multiple eutrophication mitigation interventions for example control of nutrient loading, physicochemical and physicomechanical interventions in order to take care of legacy P and ensure successful mitigation process.

Acknowledgements

Special thanks to College of Agriculture and Food Sciences, Florida A&M University for providing a conducive environment for writing this book chapter. This work was supported by USDA/NIFA (1890 Evans-Allen Research Program), Division of Research Florida A&M University and USDA-Forest Service grant number 17-CA-11330140-027.

Author details

Lucy Ngatia* and Robert Taylor

*Address all correspondence to: lucy.ngatia@famu.edu

College of Agriculture and Food Sciences, Florida A&M University, Tallahassee, FL, USA

References

[1] Xu H, Paerl HW, Qin B, Zhu G, Gao G. Nitrogen and phosphorus inputs control phytoplankton growth in eutrophic Lake Taihu, China. Limnology and Oceanography. 2010, 2010;**55**(1):420-432

[2] Nixon SW. Coastal marine eutrophication: A definition, social causes, and future concerns. Ophelia. 1995;**41**:199-219

[3] Paerl HW. Controlling eutrophication along the freshwater–marine continuum: Dual nutrient (N and P) reductions are essential. Estuaries and Coasts. 2009;**32**:593-601

[4] Bennett EM, Carpenter SR, Caraco NF. Human impact on erodable phosphorus and eutrophication: A global perspective: Increasing accumulation of phosphorus in soil threatens rivers, lakes, and coastal oceans with eutrophication. Bioscience. 2001;**51**(3):227-234

[5] Bledsoe EL, Philips EJ, Jett CE, Donnelly KA. The relationships among phytoplankton biomass, nutrient loading and hydrodynamics in an inner-shelf estuary. Ophelia. 2004;**58**:29-47

[6] Carpenter SR. Eutrophication of aquatic ecosystems: Bistability and soil phosphorus. PNA. 2005;**102**:10002-10005

[7] Pierzynski GM, Sims JT, Vance GF. Soils and Environmental Quality. Boca Raton, CRC Press; 2000

[8] Smayda TJ. Harmful algal blooms: Their ecophysiology and general relevance to phytoplankton blooms in the sea. Limnology Oceanography. 1997;**42**:1137-1153

[9] Paerl HW. Nuisance phytoplankton blooms in coastal, estuarine, and inland waters. Limnology and Oceanography. 1988;**33**:823-847

[10] Carpenter SR, Caraco NF, Correll DL, Howarth RW, Sharpley AN, Smith VH. Nonpoint pollution of surface waters with phosphorus and nitrogen. Ecological Applications. 1998;**8**: 559-568

[11] MacDonald GK, Bennett EM, Potter PA, Ramankutty N. Agronomic phosphorus imbalances across the world's croplands. Proceedings of the National Academy of Sciences of the United States of America. 2011;**108**:3086-3091

[12] Fluck RC, Fonyo C, Flaig E. Land-use-based phosphorus balances for Lake Okeechobee, Florida, drainage basins. Applied Engineering in Agriculture. 1992;**8**:813-820

[13] Stutter MI, Shand CA, George TS, Blackwell MSA, Bol R, MacKay RL, Richardson AE, Condron LM, Turner BL, Haygarth PM. Recovering phosphorus from soil: A root solution? Environmental Science & Technology. 2012;**46**:1977-1978

[14] Michalak AM, Anderson EJ, Beletsky D, Boland S, Bosch NS, Bridgeman TB, Chaffin JD, Cho K, Confesor R, Daloglu I, et al. Record-setting algal bloom in Lake Erie caused by agricultural and meteorological trends consistent with expected future conditions. Proceedings of the National Academy of Sciences of the United States of America. 2013;**110**:6243-6244

[15] Orderud GI, Vogt RD. Trans-disciplinarity required in understanding, predicting and dealing with water eutrophication. International Journal Sustainable Development World Ecology. 2013;**20**(5):404-415

[16] Schindler DW, Carpenter SR, Chapra SC, Hecky RE, Orihel DM. Reducing phosphorus to curb lake eutrophication is a success. Environmental Science & Technology. 2016;**50**:8923-8929

[17] Haygarth PM, Condron LM, Heathwaite AL, Turner BL, Harris GP. The phosphorus transfer continuum: Linking source to impact with an interdisciplinary and multi-scaled approach. Science of the Total Environment. 2005;**344**:5-14

[18] Meinikmann K, Hupfer M, Lewandowski J. Phosphorus in groundwater discharge—A potential source for Lake eutrophication. Journal of Hydrology. 2015;**524**:214-226

[19] Jaworski NA, Groffman PM, Keller AA, Prager JC. A watershed nitrogen and phosphorus balance: The upper Potomac River basin. Estuaries. 1992;**15**:83-95

[20] Attiwill PM, Adams MA. Nutrient cycling in forests. New Phytologist. 1994;**124**:561-582

[21] Turner BL. Resource partitioning for soil phosphorus: A hypothesis. Journal of Ecology. 2008;**96**:698-702

[22] Condron LM, Turner BL, Cade-Menun BJ. The chemistry and dynamics of soil organic phosphorus. In: Sims JT, Sharpley AN, editors. Phosphorus: Agriculture and the Environment. Madison, Wisconsin, USA: ASA-CSSASSSA; 2005. pp. 87-121

[23] Negassa W, Leinweber P. How does the Hedley sequential phosphorus fractionation reflect impacts of land use and management on soil phosphorus: A review. Journal of Plant Nutrition and Soil Science. 2009;**172**:305-325

[24] Ngatia LW, Hsieh YP, Nemours D, Fu R, Taylor RW. Potential phosphorus eutrophication mitigation strategy: Biochar carbon composition, thermal stability and pH influence phosphorus sorption. Chemosphere. 2017;**180**:201-211

[25] Turner BL. Organic phosphorus transfer from terrestrial to aquatic environments. In: Turner BL, Frossard E, Baldwin DS, editors. Organic Phosphorus in the Environment. Wallingford: CAB International; 2005. pp. 269-295

[26] Tarafdar JC, Claassen N. Organic phosphorus compounds as a phosphorus source for higher plants through the activity of phosphatases produced by plant roots and microorganisms. Biology and Fertility of Soils. 1988;**5**:308-312

[27] Adams MA, Pate JS. Availability of organic and inorganic forms of phosphorus to lupins (*Lupinus* spp.). Plant and Soil. 1992;**145**:107-113

[28] Zhang R, Wu F, Liu C, Fu P, Li W, Wang L, Liao H, Guo J. Characteristics of organic phosphorus fractions in different trophic sediments of lakes from the middle and lower reaches of Yangtze River region and Southwestern Plateau, China. Environmental Pollution. 2017;**152**(2):366-372

[29] De Montigny C, Prairie Y. The relative importance of biological and chemical processes in the release of phosphorus from a highly organic sediment. Hydrobiologia. 1993;**253**:141-150

[30] Bostrom B. Potential mobility of phosphorus in different types of lake sediment. Internationale Revue der Gesamten Hydrobiologie. 1984;**69**:457-475

[31] Bostrom B, Andersen JM, Fleischer S, Jansson M. Exchange of phosphorus across the sediment water interface. Hydrobiologia. 1988;**170**:229-244

[32] Redshaw CJ, Mason CF, Hayes CR, Roberts RD. Factors influencing phosphate exchange across the sediment-water interface of eutrophic reservoirs. Hydrobiologia. 1990;**192**: 233-245

[33] Perkins RG, Underwood GJC. The potential for phosphorus release across the sediment-water interface in a eutrophic reservoir dosed with ferric sulphate. Water Research. 2001;**35**(6):1399-1406

[34] Daniel TC, Sharpley AN, Edwards DR, Wedepohl R, Lemunyon JL. Minimizing surface water eutrophication from agriculture by phosphorus management. Journal of Soil and Water Conservation. 1994;**49**:30-38

[35] Sharpley AN, Chapra SC, Wedepohl R, Sims JT, Daniel TC, Reddy KR. Managing agricultural phosphorus for protection of surface waters: Issues and options. Journal of Environmental Quality. 1994;**23**:437-451

[36] Haygarth PM, Jarvis SC. Transfer of phosphorus from agricultural soils. Advances in Agronomy. 1999;**66**:195-249

[37] Edwards AC, Withers PJA. Linking phosphorus sources to impacts in different types of water body. Soil Use Manage. 2007;**23**:133-143

[38] Robertson WD. Irreversible phosphorus sorption in septic system plumes? Ground Water. 2008;**46**(1):51-60

[39] Roy JW, Robillard JM, Watson SB, Hayashi M. Non-intrusive characterization methods for wastewater-affected groundwater plumes discharging to an alpine lake. Environmental Monitoring and Assessment. 2009;**149**(1-4):201-211

[40] Wilson MA, Carpenter SR. Economic valuation of freshwater ecosystem services in the United States: 1971-1997. Ecological Applications. 1999;**9**:772-783

[41] Le C, Zha Y, Li Y, Sun D, Lu H, Yin B. Eutrophication of Lake waters in China: Cost, causes, and control. Environmental Management. 2010;**45**:662-668

[42] Ruuhijarvi J, Rask M, Vesala S, Westermark A, Olin M, Keskitalo J, Lehtovaara A. Recovery of the fish community and changes in the lower trophic levels in a eutrophic lake after a winter kill of fish. Hydrobiologia. 2010;**646**:145-158

[43] Diercks-Horn S, Metfies K, Jackel S, Medlin LK. The ALGADEC device: A semi-automated rRNA biosensor for the detection of toxic algae. Harmful Algae. 2011;**10**:395-401

[44] James KJ, Carey B, o'Halloran JO, Van Pelt FNAM, Skrabakova Z. Shellfish toxicity: Human health implications of marine algal toxins. Epidemiology and Infection. 2010;**138**:927-940

[45] Kirkpatrick B, Fleming LE, Squicciarini D, Backer LC, Clark R, Abraham W, Benson J, Cheng YS, Johnson D, Pierce R, Zaias J, Bossart G, Baden DG. Literature review of Florida red tide: Implications for human health effects. Harmful Algae. 2004;**3**(2):99-115

[46] Behm D. Coroner Cites Algae Inteen's Death. Milwaukee: JournalSentinel; 2003

[47] Weirich CA, Miller TR. Freshwater harmful algal blooms: Toxins and children's health. Current Problems in Pediatric and Adolescent Health Care. 2014;**44**:2-24

[48] Scholin CA et al. Mortality of sea lions along the Central California coast linked to a toxic diatom bloom. Nature. 2000;**403**:80-84

[49] Brooks BW, Lazorchak JM, Howard MDA, Johnson MV, Morton SL, Perkins DAK, Reavie ED, Scott GI, Smith SA, Steevens JA. Are harmful algal blooms becoming the greatest inland water quality threat to public health and aquatic ecosystems? Environmental Toxicology and Chemistry. 2016;**35**(1):6-13. DOI: 10.1002/etc.3220

[50] Smith VH. Eutrophication of freshwater and coastal marine ecosystems: A global problem. Environmental Science and Pollution Research International. 2003;**10**:126-139

[51] Schoumans OSF, Chardon WJ, Bechmann ME, Gascuel-Odoux C, Hofman G, Kronvang B, Rubaek GH, Ulén B, Dorioz JM. Mitigation options to reduce phosphorus losses from the agricultural sector and improve surface water quality: A review. Science of the Total Environment. 2014;**468-469**:1255-1266

[52] Jarvie HP, Sharpley AN, Withers JA, Scott JT, Haggard BE, Neal C. Phosphorus mitigation to control river eutrophication: Murky waters, inconvenient truths, and "Postnormal" science. Journal of Environmental Quality. 2013;**42**(2):295-304

[53] Sharpley A, Jarvie H, Buda A, May L, Spears B, Kleinman P. Phosphorus legacy: Overcoming the effects of past management practices to mitigate future water quality impairment. Journal of Environmental Quality. 2013;**42**:1308-1326

[54] Padisak J, Reynolds CS. Selection of phytoplankton associations in Lake Balaton, Hungary, in response to eutrophication and restoration measures, with special reference to cyanoprokaryotes. Hydrobiologia. 1998;**384**:41-53

[55] Qin BQ, Yang LY, Chen FZ. Mechanism and control of Lake eutrophication. Chinese Science Bulletin. 2006;**51**(19):2401-2412

[56] Yang M, Wu XG, Zhuang WH, Fang T. Application of aquatic plant in ecological restoration of eutrophic water. Environmental Science and Technology. 2007;**30**(7):98-101

[57] Kadlec RH, Wallace S. Treatment Wetlands. 2nd ed. Boca Raton, Florida: CRC Press; 2008

[58] Mitsch WJ, Gosselink JG. Wetlands. 4th ed. New York: John Wiley & Sons Inc.; 2011

[59] Vymazal J. Removal of nutrients in various types of constructed wetlands. Science of the Total Environment. 2007;**380**:48-65

Chapter 5

Phosphorus Recovery by Crystallization

Harada Hiroyuki and Inoue Katsutoshi

Additional information is available at the end of the chapter

http://dx.doi.org/10.5772/intechopen.81549

Abstract

A bubble column, a fluidized bed, and a continuous stirred tank reactor were evaluated as equipment for recovering phosphorus from various wastewaters. Magnesium was added to the solution which contained ammonia and potassium with phosphorus at high concentrations such as livestock wastewater, dehydrated water from a sewage plant, and synthetic livestock wastewater. Magnesium ammonium phosphate or magnesium potassium phosphate could be recovered by adjusting the pH of the solution. Alternatively, calcium was added and then the pH was adjusted to obtain hydroxyapatite solid conditions without seed crystals.

Keywords: magnesium ammonium phosphate, magnesium potassium, phosphate, hydroxyapatite, crystallization, phosphorus, bubble column with draft tube, fluidized bed reactor, continuous stirred tank reactor, seed recycling

1. Background

1.1. Types of crystallization for phosphorus recovery

Crystallization is considered a useful method for phosphorus recovery from various aqueous phases. The technique converts liquid phase phosphorus (usually present as a phosphate-type anion) to solid phase by utilizing the solubility products of a phosphorus-containing salt. Crystals of the relevant reactants are charged to a reactor. When the pH is adjusted to a region where the formation of new fine crystals does not occur, deposition occurs on the seed crystal. There are several dephosphorization methods, the selection of which depends on the composition of the wastewater to be treated. Whether calcium and magnesium salts are chosen depends on the relative concentrations of ammonium or potassium ions and phosphorus already present in the wastewater.

The chemical reaction for recovery of phosphorus by precipitation as magnesium ammonium phosphate (MAP, struvite) and its solubility product pK [1] is presented in Eq. (1):

$$Mg^{2+} + HPO_4^{\ 2-} + NH_4^{\ +} + 6H_2O \rightarrow MgNH_4PO_4 \cdot 6H_2O \quad pK = 13.76 \tag{1}$$

The reaction for the production of magnesium potassium phosphate (MPP) and pK [2] is presented in Eq. (2):

$$Mg^{2+} + HPO_4^{\ 2-} + K^+ + 6H_2O \rightarrow MgNH_4PO_4 \cdot 6H_2O + H_2O \; pK = 11.67 \tag{2}$$

Eq. (3) shows the reaction for the production of hydroxyapatite (HAP) [3]:

$$5Ca^{2+} + 3PO_4^{\ 3-} + OH^- \rightarrow Ca_5\,(PO_4)_3(OH) \quad pK = 57 \tag{3}$$

Phosphorus recovery by MAP and MPP crystallization required high concentration of ammonium and potassium as coexisting species, respectively, and phosphorus reacts efficiently when the solubility products are exceeded by addition of magnesium and adjustment of pH [4–6]. Wastewaters that satisfy these crystallization conditions are limited. To spread and establish this technology, it needs to be applicable to wastewaters with inadequate crystallization conditions and still obtain good results.

HAP techniques at low concentrations are well established. At high-concentration conditions, fine crystals are formed and there is concern regarding their outflow.

1.2. Reactor

In the crystallization process, stable flow conditions and a crystallizer are ideally preferred. Continuous operation makes it possible to achieve large production volumes. Continuous stirred tank reactors (CSTRs), fluidized bed reactors, and bubble column-type reactors can be employed. The basic equipment for achieving phosphorus recovery from higher concentrations consists of a stirred tank and a settling tank. Crystals in the effluent from the stirred tank become fine particles because of high supersaturation. Such particles are not easy to separate from the water treated, so the precipitate slurry is returned to the stirred tank from the bottom of the settling tank to act as seed crystal. Owing to the possibility of fine solids adhering to the return line and stirring propeller in a CSTR, the use of fluidized bed and bubble column reactors becomes preferable.

The fixed bed reactor used for low concentrations involves feeding the influent through the bottom of a rector that is filled with seed crystal and brought into contact with the seed crystal. In a fluidized bed reactor, the reaction surface area is expanded by fluidizing the seed crystal. The residence time is short, so the degree of supersaturation of the inflow is lowered by circulating treated water. In a bubble column with a draft tube, a circulating flow is generated between the inner and outer pipes by air fed to the inner tube. A sedimentation tank is provided at the upper end of the reaction section so that the crystals generated do not flow out. Seed crystals are not used.

1.3. Aqueous solutions containing phosphorus

Waste liquids containing high concentrations of phosphorus available for recovery include urine, livestock wastewater, sidestream, and sludge incineration ash extract liquid.

1.3.1. Urine

Xu et al. [4] prepared synthetic urine and examined conditions for MPP production. Potassium and potassium concentrations were about 960 and 2000 mg/L. Suitable operating pH values for MPP and MAP production were about 11 and 10, respectively. MAP treatment was also performed in a feed solution containing 240 mg/L-P and 4591 mg/L NH_4-N [5].

1.3.2. Piggery wastewater

Livestock wastewater treatment consisted of primary solid-liquid (S/L) separation and secondary biological treatment of the effluent. MAP treatment of the supernatant of primary treatment of a wastewater containing 985 mg/L NH4-N, 161 mg/L PO4-P, and 797 mg/L was reported [6].

For secondary treated water, it is advantageous for S/L separation of crystals generated in the case of membrane separation activated sludge; however, ammonia and phosphorus are reduced by biological treatment, so there are few attempts to apply MAP precipitation. One of the targets of this research is secondary treated water, particularly for operation conditions at lower concentration levels of phosphorus, ammonium, and potassium than ordinarily MAP and MPP treatments [7, 8].

1.3.3. Synthetic wastewater

Potassium chloride, phosphoric acid, magnesium chloride, and ammonium chloride were used to prepare synthetic wastewater.

1.3.4. Sidestream in sewage plant

When anaerobic digestion of excess sludge is undertaken, the concentration of phosphorus and ammonium ions in the dehydrated filtrate increases, and sufficient efficiency can be obtained using the MAP method. However, at intermediate concentrations, the efficiency of MAP decreases. Therefore, in this study, we also aimed to recover phosphorus from intermediate concentration phosphorus wastewaters. The HAP method using a CSTR reactor without seed crystal and the MAP method using a fluidized bed reactor were applied. HAP treatment was appropriate for operating conditions at higher phosphorus concentrations than ordinarily found for secondary treatment [9].

2. Phosphorus recovery characteristics using magnesium ammonium phosphate (MAP) methods from actual water with a suitable crystallizer

2.1. Magnesium ammonium phosphate treatment of secondary treatment effluent of membrane bioreactor: effects of ammonium ion concentration on phosphorus removal

In many cases of livestock wastewater treatment, the membrane separation activated sludge method is selected in preference to sewage treatment. Because the treated water does not contain suspended solids, it is not necessary to separate the recovered phosphorus and this method is therefore effective for recovering phosphorus. The precipitate of the crystallization treatments can be used as received; however, if a nitrification process is used, the ammonium concentration decreases, which is disadvantageous for MAP. Raw urine was purchased from a livestock farmer and placed in the 2 m^3 urine sump at Kumamoto Prefectural Livestock Research Institute. The solution was progressively moved into the aeration tank at 100 L/day and then treated with a tower-type membrane-activated sludge methods. The treated water was received in a 1 t tank and used as the raw water for MAP treatment experiments. **Figure 1** shows MAP experimental apparatus, which consisted of a reaction region with a draft tube structure and precipitation region for S/L separation, both volumes of 5 L.

The reaction region was 15 cm in diameter and 28 cm in height. A double pipe was installed at the bottom, through which circulating flow was generated by injection air. The hydraulic retention time was set to 10 h and activated sludge-treated water was continuously supplied to the reactor bottom. The pH was maintained at 7.5–8.5 via a pH controller dosing a 0.5 N NaOH solution and aeration at 12 L/min carried out using an air pump. The equipment was located outdoors without temperature control. The experiment was carried out for about 100 days. Sampling was conducted every 2–3 days. Changes in water properties are shown as raw and treated waters in **Figure 2**. We try to remove residual concentration by adsorption. Because the phosphorus concentration in the effluent was not adequately decreased by MAP treatment, further treatment was conducted using adsorption methods. Phosphorus-adsorbent pellets (Limonite Co. Ltd., Kumamoto, Japan) were made from a special soil. The soil was compressed to pellets of 11–13 mm diameter and 10–30 mm length with a binder. The main component of the soil was 62% iron oxide, which exhibited phosphorus adsorption characteristics. The pellets were placed in column of 30 cm height and 10 cm diameter, and MAP-treated water was supplied from the bottom at 20 L/day.

The total phosphate (T-PO_4) content of the raw water approximately 100 mg/L up to 80 days after which decreased to 80 mg/L. The treatment characteristic could be categorized into

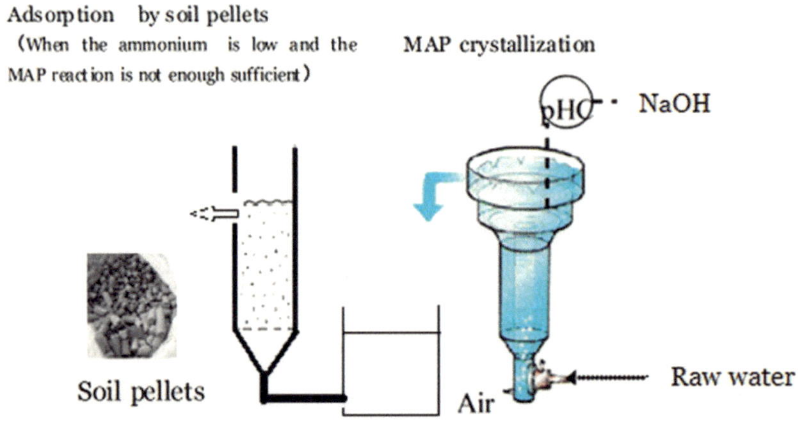

Figure 1. Bubble column with draft tube for magnesium ammonium phosphate (MAP) (or magnesium potassium phosphate, MPP) treatment.

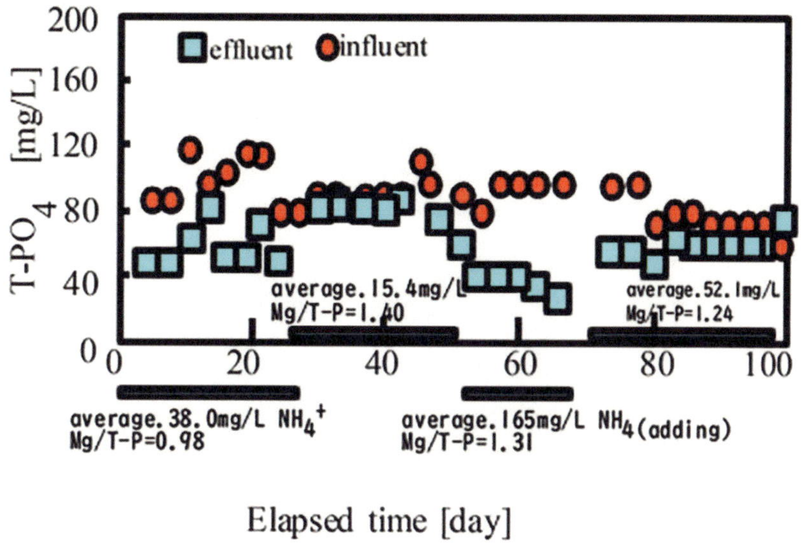

Figure 2. Magnesium ammonium phosphate treatment of secondary livestock wastewaters.

periods: 0–24, 25–54, 50–64, and 71–100 days. The average ammonium concentrations during these periods were 38, 15, 165, and 52 mg/L, respectively. The higher concentrations (165 mg/L) were due to addition of ammonium chloride. The Mg/T-P ratio was maintained at 0.98–1.53. When the raw water phosphoric acid concentration was 100 mg/L, the extent removal increased in relation to the ammonium concentration. Even if the ammonium ion concentration increased, the phosphorus removal decreased when the phosphoric acid concentration decreased to 80 mg/L. Weijia et al. [7] also considered the addition of magnesium and ammonium to phosphorus recovery using a bubble column, because the phosphorus and ammonium concentrations of the activated sludge secondary treatment are reduced by nitrification and biological uptake. The results showed about 50% phosphorus removal and that remaining treatment was required. Attempts have also been made to add magnesium and ammonium to increase the MAP production when the ammonium concentration is low [7]. The chemical reaction for recovery of phosphorus by precipitation as magnesium (MAP, struvite) is shown in **Figure 3**.

The adsorption results are shown in **Figure 4**. The phosphorus concentration of MAP-treated water was in the range 44–58 mg/L but decreased to 7–21 mg/L because of adsorption. The total average removal efficiency was 74%.

2.2. Magnesium ammonium phosphate treatment of sidestream in sewage treatment: high-speed solid-liquid separation

High concentrations of phosphorus and ammonium ions are contained in the dehydrated filtrate of the sewage, and fine crystals are precipitated unless adequate S/L separation is performed. The average total phosphorus and ammonium concentrations of the dehydrated filtrate at the North Area Kumamoto sewage treatment plant during the test period were 86.4 and 600 mg/L, respectively.

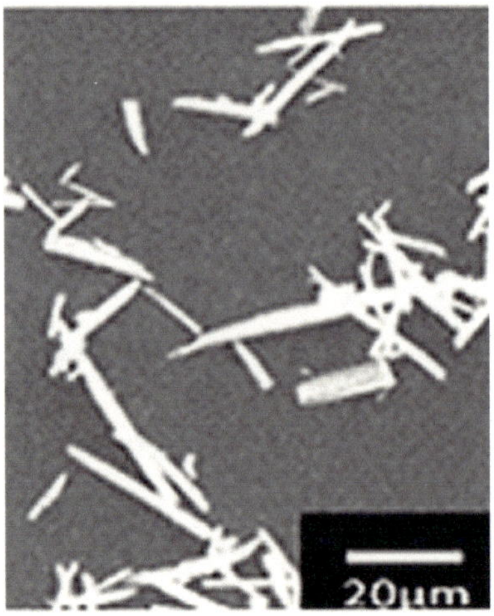

Figure 3. Magnesium ammonium phosphate crystals.

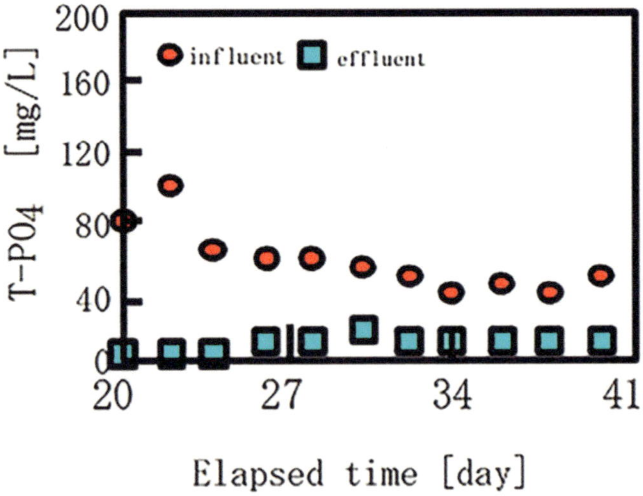

Figure 4. Adsorption treatment of magnesium ammonium phosphate effluent.

2.2.1. Fluidized bed reactor

The fluidized bed reactor apparatus is shown in **Figure 5**. The reactor was acrylic and had a diameter of 2 cm and height of 100 cm. There was no settling region.

The bottom part was filled to a height of 4 cm with glass beads with a diameter of 5 mm to ensure laminar flow. The pH of the raw was around 6.7, and there was almost no change in the phosphorus concentration on adding magnesium. To reduce pH fluctuations, part of the sidestream was diluted with raw water at 1:1. The mixed water was withdrawn and then supplied to the reactor.

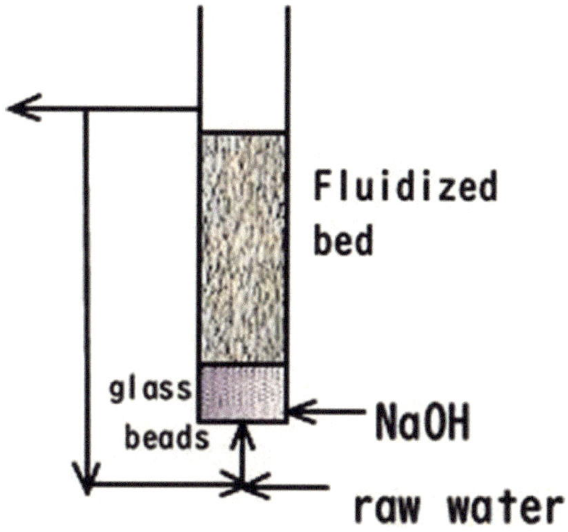

Figure 5. Fluidized bed reactor.

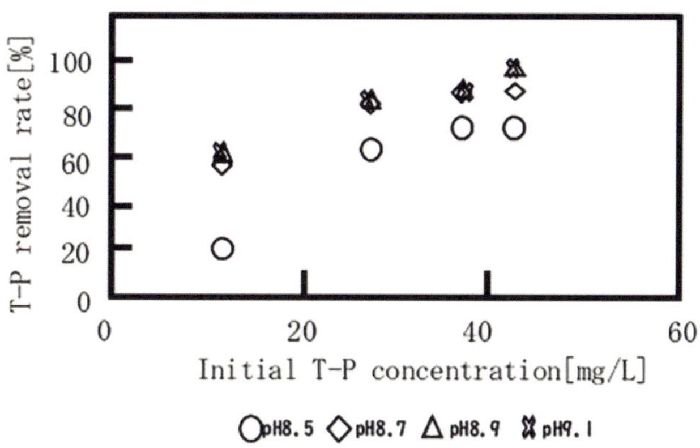

Figure 6. Relationship between initial total phosphorus (T–P) concentration and extent of its removal as a function of pH of effluent.

Solids were suspended by injecting the fluid in an upward direction. Contact between the solid and liquid was large and their superficial velocity was high. Based on these advantages, the MAP rapidly precipitated and deposited on the solid particles, enabling a very short processing time. To lower the degree of supersaturation at the inlet, circulate treated water and mix with raw water to supply. The lack of need for a sedimentation tank also contributed in processing time. **Figure 6** shows the relationship between T-P removal efficiency and pH at different initial P concentrations. At an initial concentration of 11.4 mg/L (PO_4: 34.2 mg/L), the phosphorus removal was about 20% at pH 8.5, but this increased to approximately 60% as the pH increased to 9.1. At an initial concentration 27.2 mg/L (PO_4: 81.6 mg/L), the removal reached 80%. Uludag-Demirer and Othman [8] investigated removal from wastewater containing phosphate 100–1500 mg/L and reported 48% removal at pH 9.0 for water containing

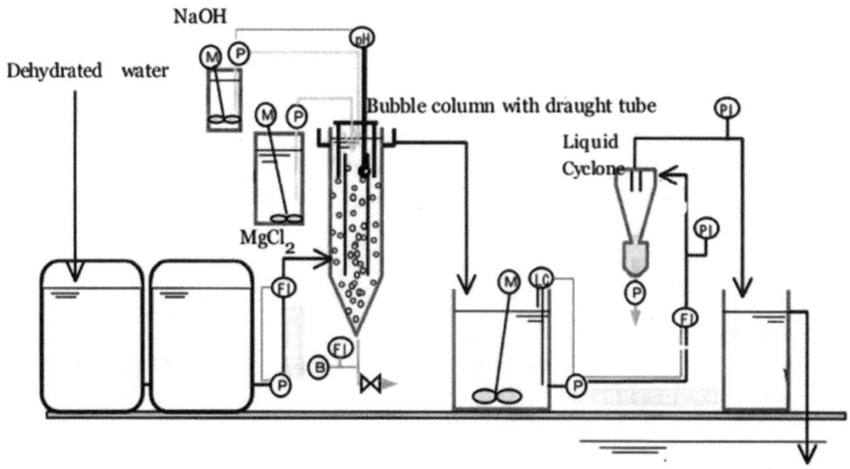

Figure 7. System configuration of mini-plant in sewage treatment center.

100 mg/L phosphate. Guadie et al. [9] also investigated for high (120 mg/L)- and low (12.5 mg/L)-concentration wastewaters. They found that T-P removal at low concentration could be improved by circulation of the treatment water. Su et al. [10] pointed out the effects of seed crystal.

2.2.2. Mini-plant experiment: combination of bubble column with draft tube and liquid cyclone

Unitika Co. Ltd. Central Research Center (Kyoto, Japan) set up a small-scale MAP reactor in a sewage treatment center (Kumamoto, Japan) and applied the technique to the dehydrated liquid of methane fermentation. An overview of the plant is shown in **Figure 7**. The reactor contained an inner cylinder and circulating flow was obtained by feeding air to the inside. A long residence time was necessary to prevent the outflow of solids, so it was not suitable for treatment of large volume. We introduced a hydro-cyclone to minimize the residence time and made the water surface area of the separation section as small as possible. The design influent rate was 5 m^3/day, and the instantaneous supply to hydro-cyclone was 2 m^3/day. The reaction volume was 86 L; the residence time in the reaction and separation regions was 25 and 1.6 min, respectively. The average TP removal efficiency was 53.3% for raw water average concentration 94.7 mg/L (maximum 105 mg/L, minimum 80.7 mg/L). When a hydro-cyclone was applied, the removal improved 77.6%; however, if a longer retention time 4.6 min was applied to the separation, the average TP removal increased to 86.7% in the reactor alone [12].

3. Phosphorus recovery characteristics using hydroxyapatite (HAP) methods from actual water

For experiments under high concentrations, actual dehydrate liquids were collected from the sewage treatment center in Kumamoto Prefecture, Japan.

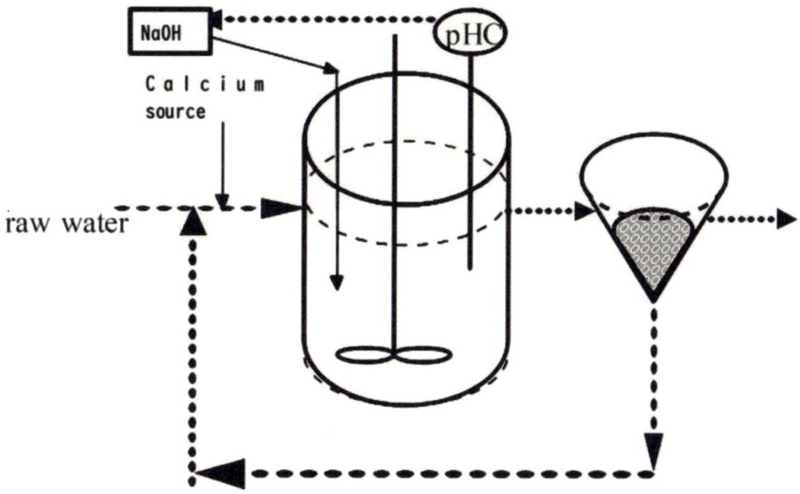

Figure 8. Reactor for hydroxyapatite treatment.

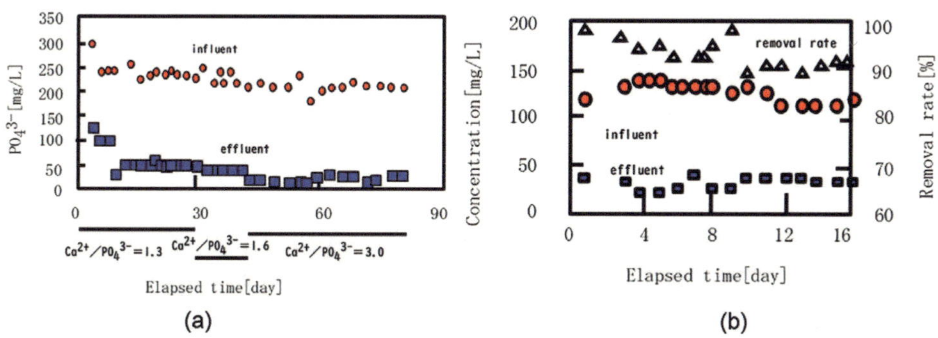

Figure 9. Results of continuous hydroxyapatite (HAP) treatment for long periods for different initial phosphorus concentrations when adding calcium chloride. In Run (a), the concentration range was 200–250 mg/L P, and the molar ratio of added calcium was varied. In Run (b), the concentration range was 100–150 mg/L P and the ratio of Ca:P was 3.0

This system has a high degree of supersaturation, so fine precipitates were easily generated, and solid-liquid (S/L) separation was difficult. Therefore, it was desirable to recover as much phosphorus as possible while promoting crystal growth. A sedimentation tank was provided and crystals of the bottom region were returned to the reactor. To examine the effects of calcium source on recovery rate, an experiment was carried out using both calcium chloride and lime (calcium hydroxide). The influence of the addition amount was examined for the calcium chloride system. The HAP apparatus shown in **Figure 8** used a 2.5 L reaction tank and 1 L settling tank. The filtrate of the digested sludge was fed to the reaction tank at 0.6 L/h and 200 mg/L calcium chloride solution was fed to maintain a Ca^{2+}/PO_4^{3-} molar ratio of 3.0. The influence of the calcium chloride addition was examined. The reflux ration from the sedimentation tank to the reaction tank was maintained at 2.0. The pH of reactor was controlled at 8.5–8.7 by adding 0.5 M sodium hydroxide. Run (a) of **Figure 9** shows the relationship

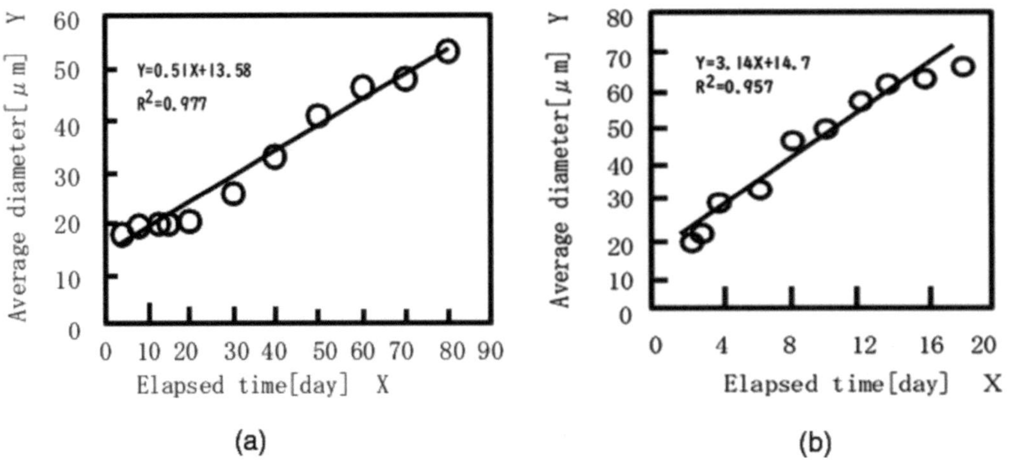

Figure 10. Crystal growth rates accompanying continuous treatment of raw water containing (a) 200–250 mg/L PO_4^{3-} (b) 100–150 mg/L PO_4^{3-}.

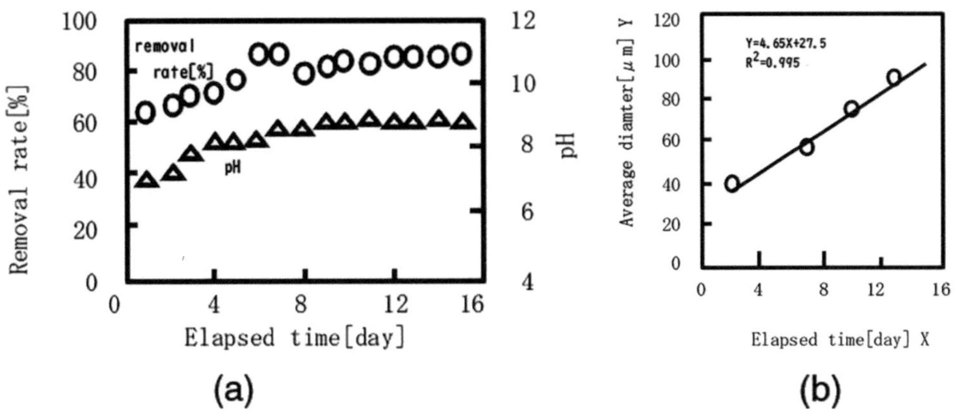

Figure 11. (a) Removal rate and (b) crystal growth during (b) to hydroxyapatite (HAP) treatment with the addition of lime.

between removal of phosphorus and calcium addition. The removal efficiency was 70–90% at Ca/PO_4^{3-} = 1.2, 75–90% at Ca/PO_4^{3-} = 1.6, and 90% or more at Ca/PO_4^{3-} = 3.0. Run (b) of **Figure 10** used a raw water concentration of about 150 mg/L PO_4^{3-}, which was two-thirds of the concentration in Run (a) because it contained wash water. When the calcium chloride addition ratio was 3.0, this gave same removal as Run (a). **Figure 10** shows changes in the average particle diameters: it took 82 days for Run (a) crystals to reach 60 μm, while this size was achieved after 12 days for Run (b). The high initial concentration of Run (a) caused the generation of a large number of primary nuclei (2). Lime feed: dehydrated filtrate was supplied at 0.8 L/h. Lime of 4000 mg/L concentration was fed to the reaction tank at 0.8 L/h and returned from the settling tank at a reflux ration of 2. **Figure 11** shows the relationship between elapsed time, pH, phosphate removal, and average diameter. The pH gradually rose over the 14-day period and reached pH 8.9. The extent of removal increased in proportion to

http://dx.doi.org/10.5772/intechopen.81549

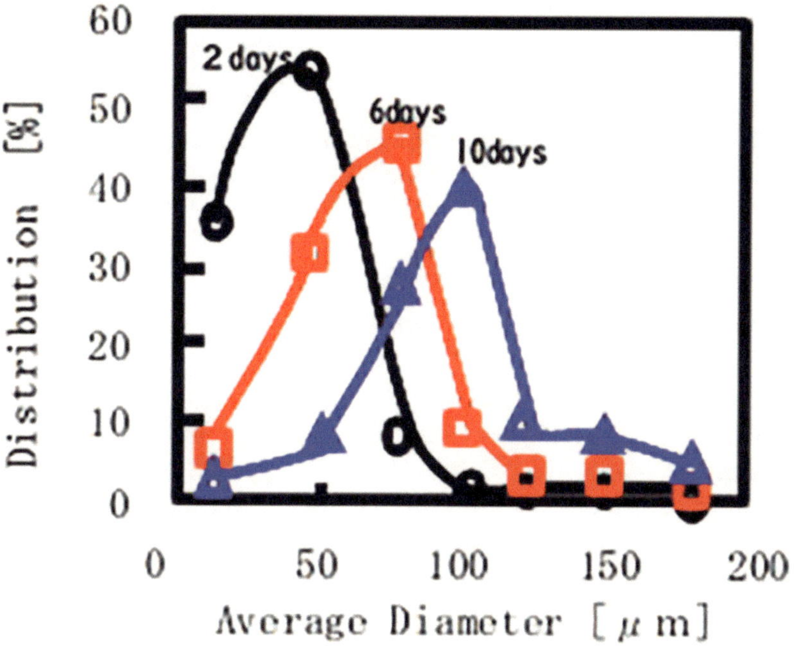

Figure 12. Distribution of diameter.

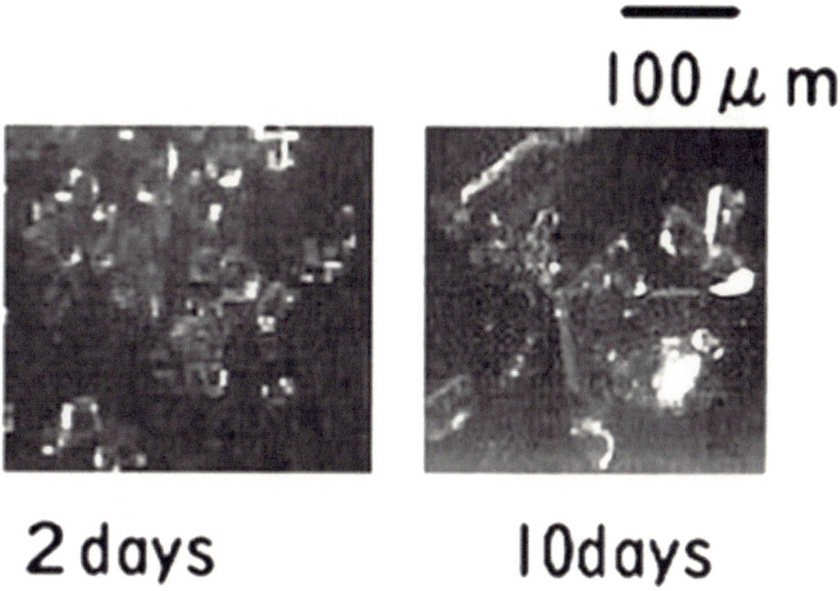

Figure 13. Photographs of crystals.

the pH, reaching approximately 90%. Changes in average particle diameter and distribution as a function of lime dosage and time are shown in **Figure 12**, and photographs are shown in **Figure 13**, respectively. When lime was present as a solid (milk of lime) and acted as a seed crystal, crystals grew rapidly. From these results, it was possible to treat actual wastewater with a phosphate concentration of 200–250 mg/L HAP crystallization. We could grow crystals by circulating precipitated crystals and retaining them for a longer period in the reactor. The appropriate ratio of calcium to phosphate was 1.6–3, and the pH was in the range of 8.5–8.7. Crystal growth was faster with a lower phosphorus concentration. The growth rate increased further with the use of lime as the calcium source.

4. A study of required conditions of magnesium potassium phosphate (MPP) precipitation by using synthetic secondly livestock wastewater

When compared with primary treatment, the phosphorus concentration also decreased because of activated sludge secondary treatment. When magnesium is added to a system in which potassium and ammonium coexist, competition arises in the formation of phosphate salts. MPP production is advantageous if the ammonium concentration is decreased in the secondary treatment. These relationships and the processing characteristics were clarified with respect to continuous processing.

This study was initially performed using batch beaker tests. To determine the influence of the stoichiometric relationship of ingredients, we used a synthetic secondary piggery wastewater (including phosphorus, potassium, and ammonium). **Figure 14** shows the effects of phosphorus concentration on removal and recovery. We had previously confirmed that the optimum pH for MPP production is pH 10–11 [12, 13]. The initial phosphorus concentrations were from 0.65 to 4.6 mM from MAP treatment; the NH_4 concentrations were set at 25.6 and 4.4 mM, respectively. Magnesium was added to be equimolar to phosphorus. Removal of phosphorus refers to a reduction in its concentration in the liquid (expressed as a percentage); recovery refers to the percentage of phosphorus reporting to the solid product. Extents of removal and recovery for an initial phosphorus concentration 0.65 mM were 37.1 and 76.2%, respectively.

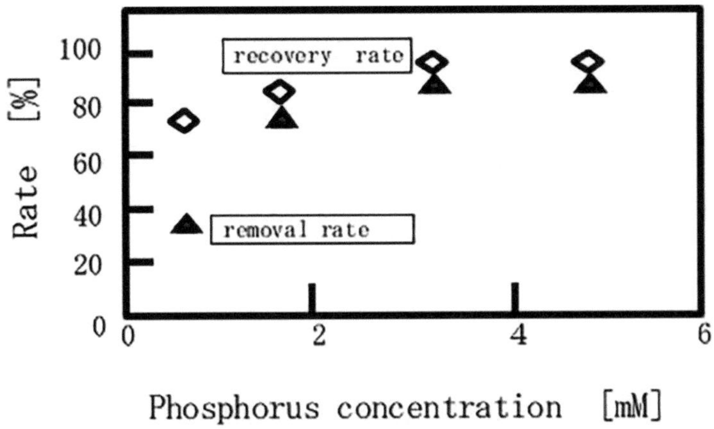

Figure 14. Changes removal rate to removal extent.

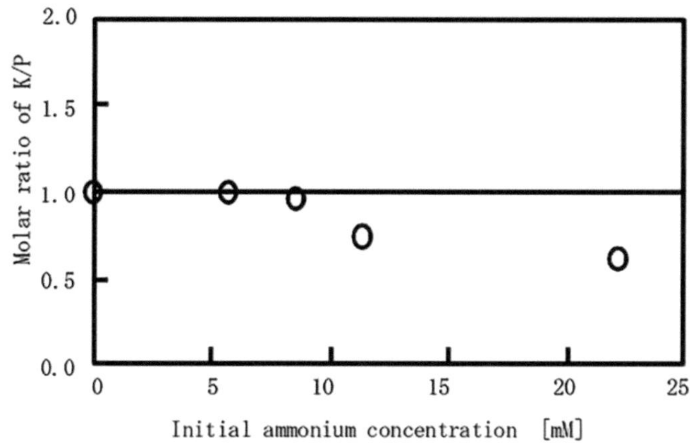

Figure 15. Effects of initial ammonium concentration on K/P [10].

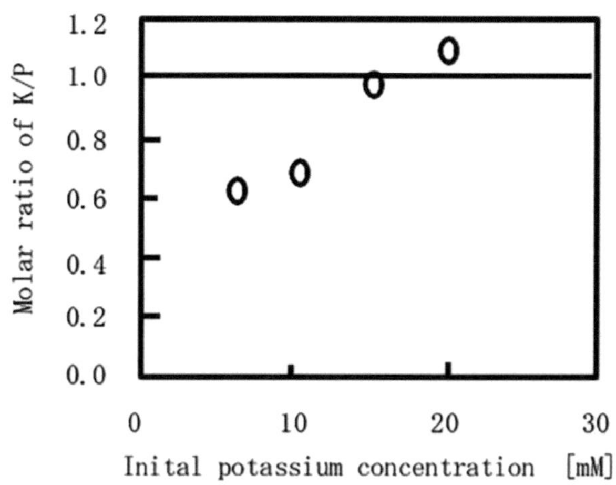

Figure 16. Effects of initial potassium concentration on K/P [10].

Both removal and recovery increased with increasing initial phosphorus concentration; more than 3.0 mM was required to reach approximately 90%. **Figure 15** shows the effects of initial ammonium concentration on the potassium to phosphorus (K:P) molar ratio in the resulting crystals. Phosphorus and potassium were 6.5 and 5.6 mM, respectively, and ammonium concentration ranged from 0 to 22.3 mM. Magnesium was added to be equimolar with phosphorus and the pH was maintained 11. The K:P molar ratio did not change to 11.6 mM Mg and then decreased, which was attributed to competition with MAP production. **Figure 16** shows the effects of potassium concentration on the molar ratio. When the potassium concentration was 17 mM (680 mg/L), the K:P ratio was closed to 1, which showed that at least 680 mg/L K is necessary for MPP production. **Figure 17** shows the influence of the additional molar ratio of magnesium on phosphorus on the component ratio in the crystal pH 11. The input of Mg:P molar ratio was changed. Phosphorus, potassium, and ammonium ion concentrations were

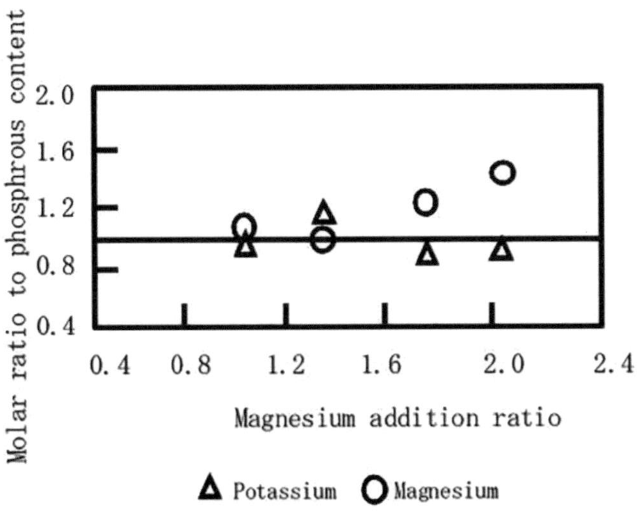

Figure 17. Relationship between the magnesium-to-phosphorus ratio in the wastewater and the potassium- and magnesium-to-phosphorus molar ratios in the resulting crystals.

Figure 18. Scanning electron microscopy analysis of white precipitation [12].

changed to 1, 1.3, 1.7, and 2. pH values were 10 and 11. When Mg:P = 1, the molar ratio of each component was a stoichiometric ratio of almost 1.

These results showed that to maintain the purity of crystals by efficient MAP treatment, it is important to maintain the pH at 11; the phosphorus and potassium and concentrations were required to be 3 mM (54 mg/L) and 17 mM (680 mg/L), respectively, and the ammonium concentration be less 8 mM (120 mg/L).

Figure 18 shows photograph of crystals obtained by continuous treatment in a bubble column using the apparatus of **Figure 1**. **Figures 19** and **20** show the result of X-ray diffraction (XRD) and energy-dispersive x-ray (EDX) analysis, respectively.

It is difficult to distinguish MAP and MPP in photographs because both crystals have a needle-like shape. Considering these results in conjunction with the XRD data, it appears that MPP was formed in Run (b), but not in Run (a). As already mentioned MAP and MPP were generated

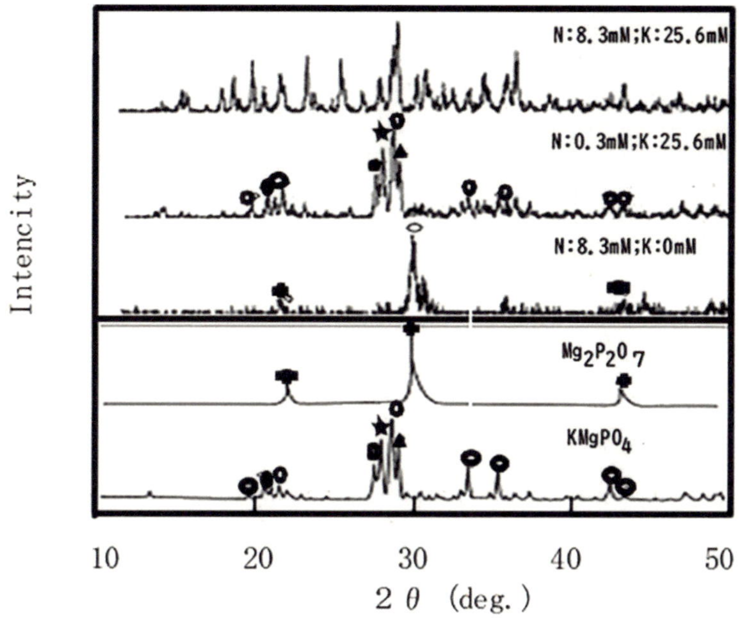

Figure 19. Comparison of X-ray diffraction patterns of sample and standard crystal [12].

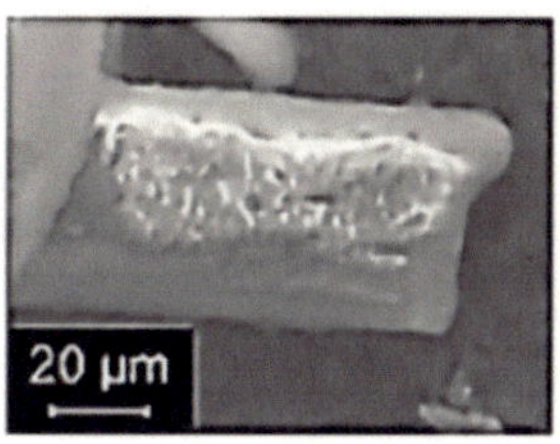

	P	O	Mg	N	K	Na
mol%	16.3	53.7	14.0	0	15.7	0.2

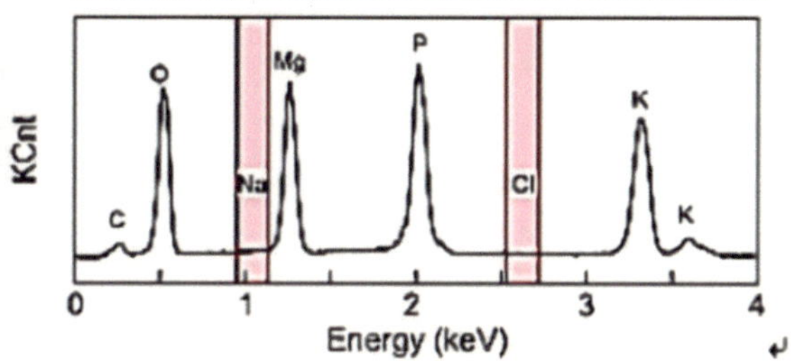

$[NH_4^+]$= 0.3 mM $[K^+]$=25.6 mM[T-P]=4.6 mM

Figure 20. Energy-dispersive X-ray analysis of white precipitate [12].

from a concentration of 8.3 mM. The XRD results of Run (a) indicated that the molar percentage ratios of P:Mg:K were 16.3:14.0:15.7, which is close to the target stoichiometry of MPP crystal.

5. Conclusion

MAP treatment for primary treated water of livestock wastewater is handled by immersing a mesh metal in liquid. In the secondary treatment, ammonium and phosphorus ion concentrations are lower by nitrification and bio-uptake, so treatment efficiency is lowered. In the membrane separation activated sludge method, because suspended solid is not mixed, recovery of phosphate is easy. To address this problem, a combination of the MAP and adsorption methods and an HAP method with crystal recirculation and without seed crystal were considered. In addition, competition with MAP precipitation is a problem in the MPP method, so inhibition of the ammonium concentration in the MPP method was studied using artificial water. MPP production was found to be given priority when the concentration of ammonium ions was 3 mM or less and potassium ions were 17 mM or more.

Author details

Harada Hiroyuki[1]* and Inoue Katsutoshi[2]

*Address all correspondence to: ho-harada@pu-hiroshima.ac.jp

1 Prefectural University of Hiroshima, Japan

2 Saga University, Japan

References

[1] Taylor AW, Frazier AW, Gurney EL. Solubility products of magnesium ammonium and magnesium potassium phosphates. Transactions of the Faraday Society. 1963;**59**: 1580-1584. DOI: 10.1039/TF9635901580

[2] Chimenos JM, Fernandez AI, Villalba G, Segarra M, Urruticoeche A, Artaz B, et al. Removal of ammonium and phosphates from wastewater resulting from the process of cochineal extraction using MgO-containing by-product. Water Research. 2003;**37**:1601-1607. DOI: 10.1016/S0043-1354(02)00526-2

[3] Bell LC, Mika H. Synthetic hydroxyapatite-solubility product and stoichiometry of dissolution. Archives of Oral Biology. 1978;**23**(5):329-336. DOI: 10.1016/0003-9969(78)90089-4

[4] Xu K, Zheng M, Zhang C, Xie T, Wang C. The precipitation of magnesium potassium phosphate hexahydrate for P and K recovery from synthetic urine. Water Research. 2015;**80**(1):71-79. DOI: 10.1016/j.watres.2015.05.026

[5] Gao Y, Liang B, Chen H, Yin P. An experimental study on the recovery of potassium (K) and phosphorus (P) from synthetic urine by crystallization of magnesium potassium phosphate. Chemical Engineering Journal. 2018;**337**:19-29. DOI: 10.1016/j.cej.2017.12.077

[6] Suzuki K, Tanaka Y, Kuroda K, Hanajima D, Fukumoto Y. Recovery phosphorus from swine waste water through crystallization. Bio/Technology. 2005;**96**:1544-1550. DOI: 10.1016/j.biortech.2004.12.017

[7] Gong W, Li Y, Luo L, Luo X, Cheng X, Liang H. Application of struvite-MAP crystallization reactor for treating cattle manure anaerobic digested slurry: Nitrogen and phosphorus recovery and crystal fertilizer efficiency in plant trials. International Journal of Environmental Research and Public Health. 2018;**15**:1397-1408. DOI: 10.3390/ijerph15071397

[8] Uludag-Demirer S, Othman M. Removal of ammonium and phosphate from the supernatant of anaerobically digested waste activated sludge by chemical precipitation. Bioresource Technology. 2009;**100**:3236-3244. DOI: 10.1016/j.biortech.2009.02.030

[9] Guadie A, Xia S, Jiang W, Zhaou L, Zhang Z, Hermanowicz SW, et al. Enhanced struvite recovery from wastewater using a novel cone-inserted fluidized bed reactor. Journal of Environmental Sciences. 2014;**26**:765-774. DOI: 10.1016/S1001-0742(13)60469-6

[10] Su C-C, Abarca RRM, de Luna MDG, Lu M-C. Phosphate recovery from fluidized-bed wastewater by struvite crystallization technology. Journal of the Taiwan Institute of Chemical Engineers. 2014;**45**(5):2395-2402. DOI: 10.1016/j.jtice.2014.04.002

[11] Harada H, Matsushita T, Fujii M, Inoue K, Kawakita H, Ohto K, et al. Evaluation of magnesium ammonium phosphate particle separate performance using combined MAP reactors with a small settling basin separate, a baffle separator at top combined with a hydrocyclone. Kagaku Kogaku Ronbunshu. 2009;**35**(1):27-29 (in Japanease)

[12] Akagi K, Teranaka R, Harada H, Katayama Y, Mitoma Y, Biswas BK. Generation of potassium phosphate magnesium from synthetic live stock wastewater containing high levels of phosphorus and potassium. Journal of Water Pollution and Treatment. 2014;**1**:1-7

[13] Satoshi Y, Seichiro O, Hiroyuki H, Kotaro A, Mitoma Y, Hidetaka K, et al. Simultaneous crystallization of phosphate and potassium as magnesium potassium phosphate using bubble column reactor with draught tube. Journal of Environmental Chemical Engineering. 2013;**1**(4):1154-1158. DOI: 10.1016/j.jece.2013.08.032

Chapter 6

Phosphorus Fertilizer: The Original and Commercial Sources

Sayma Samreen and Sharba Kausar

Additional information is available at the end of the chapter

http://dx.doi.org/10.5772/intechopen.82240

Abstract

Earlier the original source of phosphorus (P) fertilizer was bones; as time passes, the supply of P fertilizer will get exhausted. Today, rock phosphate is the only raw material in the form of P fertilizers. There are two types of rock phosphates: igneous and sedimentary; both have the same phosphate mineral, i.e., calcium phosphate of apatite group. The general formula for pure rock phosphate is $Ca_{10}(PO_4)_6(X)_2$, where X is F^-, OH^- or Cl^-. These minerals are called apatites. The most common rock phosphate mined is fluorapatite, which contains impurities like CO_3, Na and Mg. Carbonate-fluorapatite (francolite) is primary apatite mineral in the majority of phosphate rocks. The high reactivity of some phosphate rocks is due to the occurrence of francolite. The major deposits are found in the US followed by China, Morocco and Russia. The US produced about 33% of the world's rock phosphate, although nearly 50% of the world reserves are in Morocco. P fertilizers are produced from either acid-treated or heat-treated rock phosphate to break the apatite bond and to increase the water soluble P content. There are many commercially available P fertilizers like rock phosphate, phosphoric acid, calcium orthophosphates, ammonium phosphates, ammonium polyphosphate and nitric phosphates.

Keywords: phosphorus, rock phosphate, fertilizer, apatites, fluorapatite

1. Introduction

0.12% of earth crust is made up of phosphorus mineral. P is a nonrenewable natural resource present in all types of rock and soils, in all living cells, and however can form complex compounds. Mineral deposits are the major supply of phosphorus. All phosphate mineral was derived from apatite by weathering. Mostly phosphate is found in different forms like quartz, calcite, dolomite, apatite, Fe-oxide minerals and clay minerals. Apatite mineral is used for manufacturing fertilizers. Extraction of phosphorus depends on the physical properties of the rocks and its geological setting.

1.1. Historical overview

Since ancient times man used natural resources such as manures, vegetables material, and bones as fertilizers. In 1840 Liebig, the German chemist, suggested the formation of superphosphate by dissolving bones in sulfuric acid that made the P more available to plants. This practice becomes so popular that bone supply is restricted in a very short time. To overcome this problem, some workers started extraction of phosphorus from rocks; in 1847 the first commercial production of P rocks from the mining of coprolites began in Suffolk in Great Britain and peaked in 1876 when about 25,000 metric tons were mined.

1.1.1. World phosphate reserves and resources

Presently all the Phosphate reserves that are found all over the world are not "mineable" deposits, as mining of them are not economically feasible. The United States is the highest phosphate-producing country in the world, while Morocco and China are the second and third countries with respect to phosphate production. Australia and Canada are recently known sites of phosphorus mining. There are rich deposits of phosphate found in Mongolia and Peru that will fulfill the need in the future. Florida phosphate industry becomes one of the major producer and exporter of phosphate fertilizer due to good transportation and industrial infrastructure facility in America and also because a substantial layer of phosphate is only 15 to 50 feet below a soft overburden. The phosphate mining in Central Florida overshadowed other sources because of low cost of mining, large deposits and the good quality of phosphate content of Florida rock. Florida is presently providing approximately 75% of the nation's supply of phosphate fertilizer and about 25% of the world supply. In 2000, mining operations began in Ontario, Canada, of North America. Florida's phosphate is part of a deposit that stretches across the state and up the coast to the Chesapeake Bay. The phosphate mining is expanded from Central Florida to Polk and Hillsborough counties, south, to Hardee, DeSoto, Manatee and Sarasota counties. In Northern Florida phosphate deposits are present in Hamilton, Columbia and Suwannee counties.

1.1.2. Discovery of phosphate in Florida

Some three decades ago in 1880, Dr. C. A. Simmonsin of England, who owned a rock quarry for building stone in Hawthorne, sends some of his rocks to Washington, DC, for analysis. The analysis determined the presence of phosphate in the rock samples, and in 1883 he made the first attempt for mining phosphate in Florida. But it was in 1889 by Albertus Vogt and others in Marion County who began the production of the first hard rock by the Marion Phosphate Company. This was later in 1890 followed by the Dunnellon Phosphate Company, in which Vogt had ownership interest, and in this way the area was flooded by thousands of prospectors, and the great Florida phosphate boom had begun. By 1894 more than 215 phosphate mining companies were operating statewide. The boom brought wealth. But in

1900 due to consolidation and capitalization, this number had dwindled to about 50. In 1881, Captain J. Francis LeBaron, chief engineer of the US Army Corps, during his survey of Peace River of Polk County, analyzed river pebbles and confirmed the presence of phosphate, but at that time this discovery did not catch much attention. In 1886 John C. Jones and Captain W. R. McKee, of Orlando, discovered high grade phosphate along the Peace River which led to the formation of an association known as the Peace River Phosphate. Mining activity along the Peace River proceeded both in the river itself and on the adjacent land. The so-called river pebble mining was the first to be exploited. In 1888, the first shipment of Peace River phosphate pebble was launched by Arcadia Phosphate Company about a year ahead of the Peace River Phosphate Company. This phosphate discovery was kept relatively quiet. Rumors of no phosphate in Central Florida spread as a result; Polk County's phosphate deposit took a back seat the first 15 years to the hard rock region to the north. The Florida Phosphate Company and the Pharr Phosphate Company were the two phosphate mining plants found in pebble district till 1890. In 1891 Pharr started shipment of land pebble for the first time; due to this there occurred a boom in the rate of river pebble production in 1893. Phosphate mining came to North Florida in the 1960s when Occidental Petroleum Company was looking for a way to get into the fertilizer business to get profit. Occidental went north and opened a mine in White Springs where it mined phosphate until 1995, when the Potash Corporation of Saskatchewan (PCS) purchased the operation. Nowadays Mosaic and PCS Phosphate, White Springs, are the two phosphate mining companies in Florida, and the third one are US Agri-Chemicals which produce phosphate fertilizers in Central Florida.

1.1.3. Florida's phosphate deposits

A blanket of phosphate deposits covers much of the Peninsular Florida has a large phosphate deposits which consists of approximately equal parts phosphate rock, clay and sand, averages 12 to 15 feet in thickness. The matrix is buried beneath a soil that is 15–30 feet deep. By the end of 1999, approximately 300,000 acres of land, or more than 460 square miles, had been mined in Florida. Polk County is the heart of the Bone Valley mining region, and the mineable deposit in this area stretches to Hillsborough, Hardee, Manatee, and DeSoto counties. The large depositions were also found in mining in North Florida's Hamilton County from a mineable area that extends into Columbia and Suwanee counties. Similar deposition is found in both the areas. Mining in Central Florida has been moving south. As sites mine out, the draglines move to where the contiguous deposit of phosphate pebble is found. Toward the south the quality of rock decreases which brings technological challenges for the mining industries. During the past years, mining is slowed down in Polk County's southern fringe. In 2000 closing of IMC Clear Springs and Noralyn mines conveyed a close to active. Currently phosphate mining companies has opened new mining sites in Manatee, DeSoto and Hardee counties.

1.1.4. Florida phosphate products today include

1.1.4.1. Diammonium phosphate (DAP)

The fertilizer that quickly became the item of commerce as most widely used by the growers today, and it had the highest concentration of phosphate and nitrogen at 18 N–46P_2O_5–0K_2O.

1.1.4.2. Monoammonium phosphate (MAP)

This fertilizer is essentially the same as DAP, but it has a lower concentration of nitrogen at 11 N–52P_2O_5–0K_2O. It is completely water soluble and has granular material; it mixes well and frequently serves as an ingredient in bulk-blended fertilizers.

1.1.4.3. Granulated triple superphosphate (GTSP)

It is very similar to the superphosphate fertilizer that provides 46% P_2O_5, some calcium and sulfur to plants. GTSP is formed by reaction of phosphate rock with phosphoric acid.

1.1.4.4. Superphosphoric acid

It's an acid used to make a concentrated or fluid fertilizer. PCS is the acid produced only by Florida Company in North Florida.

1.1.4.5. Phosphoric acid

Phosphoric acid is used in granulation plants where ammonia is added to phosphoric acid to produce the ammoniated phosphate fertilizer. Purified food-grade phosphoric acid is used in making soft drinks.

1.1.4.6. Animal feed supplements

Defluorinated phosphate rock or phosphoric acid is used to make animal feed supplements by combining phosphate rock with phosphoric acid, sodium carbonate and then calcine or react it with lime to get dicalcium phosphate.

Sulfuric acid

This acid is used to produce phosphoric acid after reacting with phosphate rocks at phosphate plants.

1.2. World resources

Sedimentary marine phosphorites are the principle resources of phosphate rock. The world's largest sedimentary reservoirs are found in North Africa, China, the Middle East, and the United States. Valuable igneous sedimentary reservoirs are also found in Brazil, Canada, Finland, Russia, and South Africa. Substantially large phosphate deposits have been spotted near the Atlantic Ocean and the Pacific Ocean shown in **Table 1** and **Figure 1**. World resources of phosphate rock are more than 300 billion tons. There are no imminent shortages of phosphate rock.

1.2.1. History of the development of phosphorus acid and its salts as agricultural chemicals

It was observed that applications of H_3PO_3 and phosphite (Phi) were less effective as compared to phosphoric acid (H_3PO_4) and its derivatives on the first crop. With increasing rates of phosphite (Phi), phytotoxic effects were detected on the crop yield. However, nutritive role of

Countries	Mine production	
	2015	2016
United States	27,400	27,800
Algeria	1400	1500
Australia	2500	2500
Brazil	6100	6500
China	120,000	138,000
Egypt	5500	5500
India	1500	1500
Israel	3540	3500
Jordan	8340	8300
Kazakhstan	1840	1800
Mexico	1680	1700
Morocco and Western Sahara	29,000	30,000
Peru	3880	4000
Russia	11,600	11,600
Saudi Arabia	4000	4000
Senegal	1240	1250
South Africa	1980	1700
Syria	750	—
Togo	1100	900
Tunisia	2800	3500
Vietnam	2500	2800
Other countries	2470	2410
World total (rounded)	**241,000**	**261,000**

Table 1. Showing world production of phosphate in 2015 and 2016.

Phi in growth response was evident when compared to the zero-P control. Whereas researchers found Phi and H_3PO_3 treatments beneficial to the second crop, this was due to probable conversion of Phi to phosphate in the soil. In general, better yield was obtained when Phi materials were used on soils with limestone. Further scientific studies related with the significance of H_3PO_3 and its salts in agriculture did not occur for nearly 30 years, but rather their performance against plant diseases was mentioned [1]. During the disease control analysis, many incidents related to the plant's physical and chemical mobility were observed when the plants were treated with H_3PO_3 or its salts in the absence of plant pathogens, some of which are described below. Ouimette and Coffey [2] reported that the Phi were more readily absorbed into plant tissues than phosphates—very important in crops with leaf surfaces that resist foliar spray uptake. In a comprehensive review given by Guest and Grant [3] related

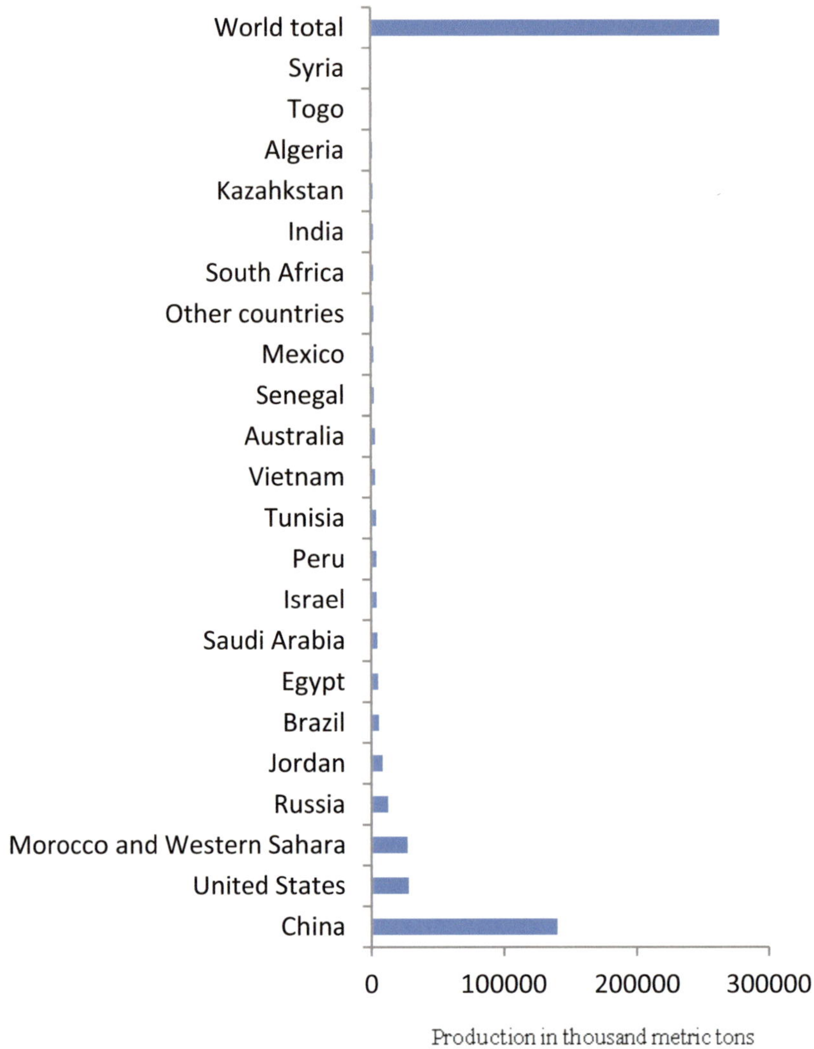

Figure 1. Phosphate rock production worldwide in 2017, by country (in 1000 metric tons).

with the complex action of phosphonates, several unique features of this chemical group were recounted. For example, Phi is a rapidly absorbing nutrient, which translocates from xylem to phloem according to normal source-sink relationships for nutrient element materials. Guest and Grant [3] reported that the Phi is more persistent as it metabolized slowly in plant tissue as compared with phosphate and does not participate in all the same biochemical pathways as phosphate. Adam and Conrad and Casida [4, 5] confirmed their results through experiments where bacterium *Pseudomonas fluorescens* 195 showed the ability to oxidize Phi and also discharge it in the growth medium as phosphate. Malacinski and Konetzka [6] repeated the same work and reported that a short adaptive period was required before oxidation of Phi by organisms, and this whole process took 14–15 weeks. Bezuidenhout et al. [7] during their study reported first time that the Phi can also be converted microbially to phosphate within plant tissues and identified three genera of bacteria (*Alcaligenes, Pseudomonas* and *Serratia*). These findings complemented the previous observations given by Rothbaum [8] that elemental P in soil was oxidized non-enzymatically under particular temperature and water. Busman et al. [9] reported that the phosphate fertilizer applied to the soil will not be utilized by the crop in the first season. Rothbaum and Baillie [8, 10] observed that Phi was less adsorbed than phosphate by the same soil. This lower 'phosphate fixation' improved growth of Phi-treated soil, with a period gap, as compared to phosphate-treated soil. Rhone-Poulenc Ag Company of the United States expressed concern to the US Environmental Protection Agency (EPA) about classifying a fungicide based on H_3PO_3 salts as a biochemical pesticide and affirmed the non-enzymatic oxidation of Phi to phosphate occur naturally over time. Lovatt [11] discovered that foliar application of K_3PO_3 to P-deficient citrus seedlings restored plant growth. This demonstrated that through metabolic processes, Phi was readily taken up by plant leaves and replaced phosphate as a source of Frazier and Waerstad [12] tested the composition and solubility of Phi to analyze the potential of this class of materials for increasing the plant nutrient element content of liquid fertilizers. Albrigo [13] reported the positive response of Phi on winter prebloom foliar of Valencia oranges which were increased flower number, fruit set and yield, plus increased total soluble solids. Additional studies by Lovatt [14] on foliar fertilization of citrus showed that application of K_3PO_3 in May and July to navel orange significantly increased the number of large-size fruit, total soluble solids and the ratio of soluble solids to acid, compared to control fruit. Biagro Western Sales, Inc., Visalia, CA, took the lead in commercialization of Phi-supplying fertilizer products patented by the University of California Anon and Lovatt [15, 16]. Today farmers are well educated and formed community of producers; they analyze themselves the effect of new Phi products on both soil and crop. In a practical sense, acceptance by discriminating growers is strong evidence that the benefits of H_3PO_3-derived fertilizers are standing up to their ultimate test—the real world of agricultural crop production.

1.2.2. Rock phosphate: its availability and solubilization in the soil

Rock phosphate is one of the basic raw materials needed in the manufacture of phosphatic fertilizers like single superphosphate, diammonium phosphate, nitrophosphates, etc. Commercial rock phosphate occurs in nature as deposits of apatites (bearing minerals) along with other accessory minerals such as quartz, silicates, carbonates, sulfates, sesquioxides, etc. Four types of rock phosphate minerals are carbonate apatite [$3Ca_3(PO_4)^2.CaCO_3$], fluorapatite [$3Ca_3 (PO_4)_2.CaF_2$], hydroxyapatite [$3Ca_3(PO_4)_2.Ca(OH)_2$], and sulpho apatite [$3Ca_3 (PO_4)_2.CaSO_4$]. Because of their well-developed crystalline formation property, the apatites of igneous and metamorphic origin are generally regarded as less reactive. However, the apatites of sedimentary rock deposits are soft minerals possessing microcrystalline structure and are of major commercial importance for direct application in the soil [17].

1.2.3. Grades of rock phosphate

The classification of reserves of indigenous rock phosphate as done by the Indian Bureau of Mines, and the purpose for which each grade can be used is given in **Table 2**.

Grade	P_2O_5 (%)	Reserve (mt)	Remarks
High	+30	15.27	Considered for wet production of fertilizers
Medium	25–30	18.95	Considered mainly for partially acid rock phosphate and for processed phosphates after less beneficiation
Low	11–25	55.22	Approx. 20% P_2O_5 grade and relatively more reactive material may be considered for partially acidulated rock phosphate production and others for direct application
Unclassified		170.04	
Total		259.48	

Table 2. Classification of known reserves of indigenous rock phosphate in India.

1.2.4. Availability of rock phosphate

Including all grades and types of rock phosphate, the known global resources are in the order of 163,000 million tons. Though globally adequate, rock phosphate is inequitably geographically distributed. Africa holds about 41%, the United States has 21%, former USSR 13%, the Middle East 10%, Asia 8%, and South America 3%, while Australia, New Zealand and Oceania together reported for only 2% and Europe >1%. Phosphate rock resources in India is, however, not very comfortable as it possesses a resource of only 260 million tons (0.19% of the world) of rock phosphate of all types and grades, catering the agricultural needs of 1/6 of the population of the world. Out of the total rock phosphate resource, the country has a predominance of low grade rock phosphate having only 15.27 million tons reserve of high grade rock phosphate (**Table 1**), and the remaining low grade rock phosphate is unacceptable to P fertilizer industry due to its very low P_2O_5 and high $CaCO_3$ content [18]. The current annual domestic demand of high grade rock phosphate is of the order of 4 million tons. Out of which 95% is consumed in agriculture sector as a source of P fertilizer. The domestic production of about 1.4 million tons/year of rock phosphate could hardly meet 35% of the total demand, and the remaining (65%) demand is met through imports. P fertilizer industry largely depends on sulfur, phosphoric acid, and ammonia besides rock phosphate. India imports around 1.7 million tons of sulfur, 2–4 million tons of phosphoric acid, 1.5 million tons of NH_3 and 4.9 million tons of rock phosphate for phosphate industry which constitutes a substantial part of our international trade in fertilizer raw material. Thus, the rapidly increasing price of soluble phosphatic fertilizer has raised interest in cheaper alternatives. Under such conditions, we must explore new methodologies for the utilization of indigenous low grade rock phosphate by converting it into a potential resource of P for direct application to the soil. The direct utilization of indigenous rock phosphate deposits could only alleviate the dependence of the country on foreign suppliers.

1.2.5. Factors affecting rock phosphate dissolution

Soils has an eminent reserve of total P, but very little amount of P is actually available to the plants to support their growth to fulfill the requirement; continuous application of phosphate fertilizers is essential for increasing crop yield. Water solubility of phosphate fertilizers depends on both acidic and neutral to alkaline conditions. Several factors that influence the application of rock phosphate as P fertilizer are rate of dissolution, soil characteristics, plant species and fertilizer.

1.2.6. Dissolution

Factors which determine rock phosphate dissolution rate are its lattice composition, accessory mineral type and particle size. Solubility of apatites increased by substituting CO_3^{2-} for PO_4^{3-} in the lattice structure due to decrease a-dimension of the unit cell, and crystal instability [19]. Silverman et al. [20] reported Calcium carbonate as the soluble apatite as compared to other apatites. Rate of its dissolution increases with the concentration of Ca^{2+} and pH at the surface of apatite, and therefore it reduces the rate of rock phosphate dissolution in soil. The rate of dissolution reduced under field conditions due to leaching or plant uptake of calcium ions. The rate of dissolution increases as the particle size decreases; this might be because fine particle size has greater degree of contact between rock phosphate and soil.

1.2.7. Soil properties affecting rock phosphate dissolution

The rate of dissolution of rock phosphate also depends on the chemical properties and type of soil to which it is applied. As compared to other parameters, pH buffering capacity was very important in soil. Earlier studies indicated that the amount of rock phosphate-P decreased with the increase in soil pH. The rate of dissolution of rock phosphate was highly sensitive to Ca^{2+} activity in the soil solution. A linear relationship between the log of Ca^{2+} activity and log of P in soil solution has been reported by Robinson and Syers [21]. Phosphate retention capacity and soil moisture also affect rock phosphate dissolution of the soil to retain P. Wet soil enhances the rate of phosphate dissolution by allowing the dissolution products. The product transported away from the surface of the rock phosphate particles and recognized the positive effect of organic matter on rock phosphate dissolution.

1.2.8. Influence of plant species on rock phosphate dissolution

Ability of plants to extract P from rock phosphates was recognized by Merril, quoted in Flach et al. [22]. Plants affect the dissolution by the secreting acid or alkali, through Ca uptake, production of chelating organic acids such as citric, malic and 2-ketogluconic acids which complex Ca and deplete P in the soil. Roots of the plants induced change in rhizosphere pH which causes imbalance in the proportion of anionic (usually NO_3–, H_2PO_4–, SO_4^{2-} and Cl^-) and cationic nutrient (K^+, Ca^{2+}, Mg^{2+} and Na^+) uptake by the plants. The imbalance in the rhizosphere is maintained by the release of either H^+ or OH^-/HCO^{3-}, thus balancing the pH of the rhizosphere. Acidic soil enhances the rate of rock phosphate dissolution. Effective rock phosphate utilization by plant species such as e.g., buckwheat and rapeseed has been responsible for their high Ca uptake. Flach et al. [22] assessed the ability of maize, pearl millet and finger millet to utilize P from rock phosphates by a pot experiment and concluded that plant species influence P dissolution; therefore, choice of crop is very important to maximize the solubility of rock phosphate.

2. Phosphorus scarcity

Today growing extraction and consumption of phosphate is exhausting existing deposits, and therefore the rate of P reserves depleting. This means that at a certain point time comes when all the phosphorus reached to the alarming peak and that condition is called 'peak phosphorus' according to literatures. This condition will be calculated on the basis of phosphate rock reserves. Since no consensus was there on the size of these reserves, so nobody knows when will be the peak phosphorus stage will occur. Peak phosphorus has been calculated by the Global Phosphorus Research Initiative (GPRI). In 2009 the GPRI estimated that phosphorus production would peak around the year 2033 and that afterward production will continuously decrease until reserves are depleted within the next 50-100 years. The US Geological Survey (USGS) re-estimated reserves at 60,000 mmt up from previous estimates of 16,000 mmt; the IFDC stated that 'there is no indication of "peak phosphorus" event within the next 20–25 years. The concept of peak phosphorus itself is contested; the main fault in the calculations of peak phosphorus is based on phosphate rock reserves not resources which provide the basis for estimation of static ranges. Phosphate rock reserve data explained by national geological surveys do not point out the absolute quantity of an element which is available for extraction, as the static paradigm would suggest. According to the 'dynamic adaptive paradigm', due to changes in economic feasibility, scarcity in the production of phosphate rock occurs. This paradigm led to scarcity which is a permanent feature of human existence: minerals become scarce as long as they are immensely valued in the society, and how much time and effort it takes to extract them, and they are related to all other goods and services in the society. Shortage of phosphate rock is an important issue when observed from a different angle that is other than relative availability. One of the reasons of fall in phosphate rock exports is the geopolitical turmoil in supplier regions. Scarcity may also result by lack of water available to the mining industry. Price inelasticity of supply, time and investments are the limiting factors which can lead to scarcity of phosphorous rock.

3. Phosphorus sustainability

Nowadays the good news is that crisis can be averted. Almost 4/5 of the phosphorus mined for food production never actually reaches the food on our forks. We can therefore invest in renewable phosphate fertilizers or innovations in on-farm efficiency to safeguard our farmers, our agriculture and food consumers. In every sector recycling of phosphorous is efficiently taking place from agriculture and mining to sanitation sector to changing diet (**Figure 2**) [23]. To meet long-term phosphorous demand of society, we have to face the technical and institutional challenges for implementing practical solutions. An integrated, context-specific approach should be developed over partial measures. Technologies and practices with effective policy instruments (regulatory, economic, facilitation) are required to encourage and bear such measures.

Food demand is rising globally with no slowing down in sight. Especially in China and other rapidly growing economies, more demand for meat and dairy means more demand for

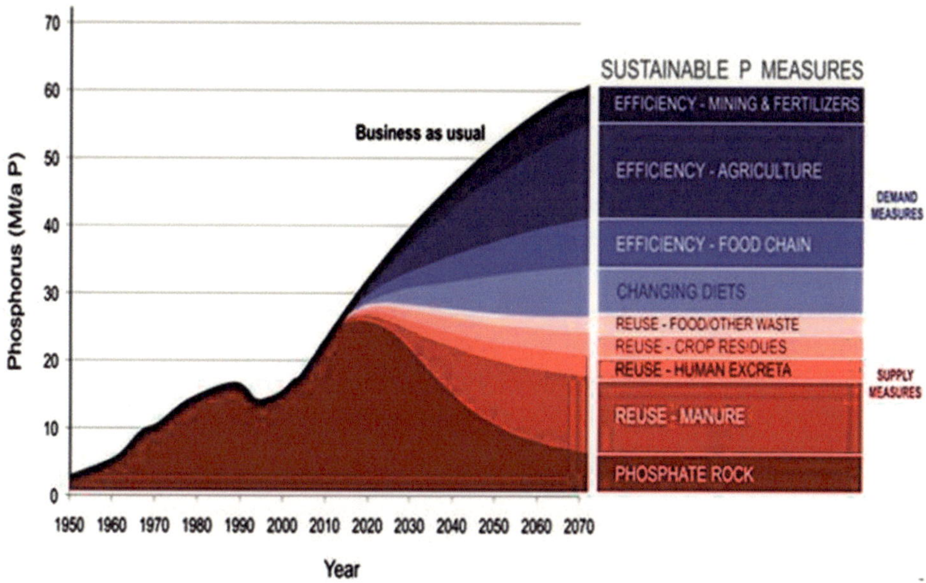

Figure 2. Sustainable phosphorus measures: Efficiency, recycling and changing diets.

fertilizers, while human body only needs around 0.4 kg of P each year. 22.5 kg of phosphate rocks are mined to meet the requirement of phosphate for each person's diet. For growing population water and energy are considered as critical for meeting future demands of food for increasing population. However, there is no approximated value of phosphorus scarcity in future as a limiting factor. Thus far we can say that without phosphorus, there would be no food and life on earth. There is no single international body responsible for managing global P resources currently in the long term, unlike oil, water and nitrogen.

4. Phosphorus recovery

Phosphorus can be recovered and used over and over again if present in sufficient concentrations dissimilar to oil, which is lost once it is used. Between the phosphate reserves and the food which we eat, up to 80% of P is lost in the process from production of fertilizers, application of phosphate on fields, in food processing and final consumption. With the increasing efficiency of phosphorous, we have to carry on recovery process of P from residues of crops, waste food items in dumpsters, manure, human excreta, struvite and other sources such as bone meal, ash and algae. A key opportunity to meeting the goal of global food security lies in the often overlooked link between addressing hunger and sanitation. In agriculture P plays a critical role as a nutrient, and but on the other hand, it is also considered as an environmental pollutant due to sewage emerging from human settlements. Human activities produce 3 million tons of P each year. If this P is recirculated back in agriculture fields from where they first came, we can maintain balance in sustaining food production in the decoupled communities which are dependent on globalized P fertilizer markets.

5. Conclusions

It may be concluded that evolution of phosphorous acid (H_3PO_3) and its salts as fertilizers owes much to both the early investigators searching for phosphate replacements and to the many scientists who later sorted out the relevant facts about plant response to Phi through phytopathological research. In recent years, scientific aroused their interest in the nutrient properties of H_3PO_3-derived products which stimulate their commercialization of H_3PO_3 as fertilizers. Trails of these fertilizers on crops have given fruitful results. The nutritive properties of Phi products proved to be a useful addition to producers' resources. Phi fertilizers elicit positive responses to crop like it enhances flowering and fruit set in cirrus and are converted to phosphate through oxidation process. H_3PO_3 derivatives give similar responses as that of orthophosphate in fertilization of crops, despite sometimes delayed. Phosphorus acid-derived fertilizers provide a more readily available source of P than that which occurs in soil. Phi products provide more phosphorus to plants as compared to phosphate fertilizers due to their high efficacy of phosphorus uptake through plant foliage. Earlier concept of not using phi as P fertilizers is now changed due to positive results recorded by the scientist about the use of these products as fertilizers. Thus we can reduce a lot of financial burden from our economy by reducing the import of rock phosphate and other by-products required in the manufacture of commercial P fertilizer.

Acknowledgements

The authors are grateful to section of Ecology and Environmental Botany, Department of Botany, A.M.U, Aligarh.

Author details

Sayma Samreen[1]* and Sharba Kausar[2]

*Address all correspondence to: samreensayma@gmail.com

1 Section of Ecology and Environmental Botany, Department of Botany, Aligarh Muslim University, Aligarh, India

2 Department of Zoology, Aligarh Muslim University, Aligarh, India

References

[1] MacIntire WH, Winterberg SH, Hardin LJ, Sterges AJ, Clements LB. Fertilizer evaluation of certain phosphorus, phosphorous, and phosphoric materials by means of pot cultures. Agronomy Journal. 1950;**42**:543-549

[2] Ouimette DG, Coffey M. Phosphonate levels in avocado (*Persea americana*) seedlings and soil following treatment with fosetyl-Al or potassium phosphonate. Plant Disease. 1989;**73**:212-215

[3] Guest D, Grant B. The complex action of phosphonates as antifungal agents. Biological Reviews. 1991;**66**:159-187

[4] Adam F, Conrad JP. Transition of phosphate to phosphate in soils. Soil Science. 1953;**75**:361

[5] Casida LE. Microbial oxidation and utilization of orthophosphite during growth. Journal of Bacteriology. 1960;**80**:237-241

[6] Malacinski G, Konetzka WA. Bacterial oxidation of orthophosphite. Journal of Bacteriology. 1966;**91**:578-582

[7] Bezuidenhout JJ, Darvas JM, Kotze JM. The dynamics and distribution of phosphite in avocado trees treated with phosetyl-Al. South African Avocado Growers' Association Yearbook. 1987;**10**:101-103

[8] Rothbaum HP, Baillie WJH. The use of red phosphorus as a fertilizer. Part 4. Phosphite and phosphate retention in soil. New Zealand Journal of Science. 1964;**7**:446-451

[9] Busman L, Lamb J, Randall G, Rehm G, Schmitt M. The Nature of Phosphorus in Soils [Internet]. 1997. URL http://www.extension.umn.edu/Documents/D/C/ DC6795. html. Communication and Educational Technology Services, University of Minnesota Extension Service; St. Paul, MN

[10] Larochelle DA. Letter to OPP/USEPA with Attachment by Helfant L, Castellino S. Dated July 18, 1996. Rhone-Poulenc Ag Company; Research Triangle Park, NC

[11] Lovatt CJ. Foliar phosphorus fertilization of citrus by foliar application of phosphite. In: Citrus Research Advisory Committee (ed.). Summary of Cirtus Research. Rierside, USA: University of California; 1992. pp. 25-26

[12] Frazier AW, Waerstad KR. Crystallography and equilibrium solubility for ammonium and potassium orthophosphites and hypophosphites. Fertilizer Research. 1992;**32**:161-168

[13] Albrigo LG. Foliar application of major elements for flowering and fruit set. Indian River Citrus Seminar, 50th Anniversary. 1997

[14] Lovatt CJ. Managing yield with foliar fertilization. Citrograph. 1998;**84**:8-13

[15] Anon. Nutri-Phite P soil; Nutri-Phite P+K; and Nutri-Phite P foliar. In: Nutri-Phite? Visalia, CA: Fertilizer Product Labels. Biagro Western Sales, Inc.; 1996

[16] Lovatt CJ. Formulation of phosphorus fertilizer for plants. U.S. Patent No. 5, 514, 200. 1996

[17] Biswas DR, Narayanasamy G. Research note: Mobilization of phosphorus from rock phosphate through composting using crop residue. Fertilizers News. 2002;**47**(3):53-56

[18] McLean EO, Ssali N. Effect of phosphorus rate and form with combination with lime and gypsum on yield and composition of German millet and alfalfa form highly weathered soil. Soil Science. 1977;**123**:155-166

[19] Lehr JR, McClellan GH. TVA Bulletine. Y-43. Muscle Shoals, Albama: Tennessee Valley Authority (TVA); 1972

[20] Silverman SR, Ruth KF, Jeanne DW. Quantitative determination of calcite associated with carbonate bearing apatites. U.S. Geological Survey; 1951. pp. 211-222

[21] Robinson JS, Syers JK. Effects of solution calcium concentration and calcium sink size on the dissolution of Gafsa phosphate rock in soils. Journal of Soil Science. 1991;**42**:389-397

[22] Flach EN, Quak W, Diest AV. A comparison of the rock phosphate-mobilizing capacities of various crop species. Tropical Agriculture. 1987;**64**:347-352

[23] Cordell D, White S. Sustainable phosphorus measures: Strategies and technologies for achieving phosphorus security. Agronomy. 2013;**3**(1):86-116. http://www.mdpi.com/2073-4395/3/1/86